OVERSIZE

ACPL ITEM
DISCARDED

553.5
Sa5s
Sanders, Scott Russell
Stone country

5008220

DO NOT REMOVE
CARDS FROM POCKET

ALLEN COUNTY PUBLIC LIBRARY

FORT WAYNE, INDIANA 46802

You may return this book to any agency, branch,
or bookmobile of the Allen County Public Library.

DEMCO

STONE COUNTRY

STONE COUNTRY

Text by Scott R. Sanders

Photographs by Jeffrey A. Wolin

INDIANA UNIVERSITY PRESS • Bloomington

This book was brought to publication with the assistance of a
grant from the Andrew W. Mellon Foundation.

Text copyright © 1985 by Scott Russell Sanders
Photographs copyright © 1985 by Jeffrey A. Wolin

All rights reserved

No part of this book may be reproduced or utilized in any form
or by any means, electronic or mechanical, including photocopying
and recording, or by any information storage and retrieval system,
without permission in writing from the publisher. The Association
of American University Presses' Resolution on Permissions constitutes
the only exception to this prohibition.

Manufactured in the United States of America

Library of Congress Cataloging in Publication Data

Sanders, Scott R. (Scott Russell), 1945–
  Stone country.

  1. Limestone—Indiana. 2. Quarries and quarrying—
Indiana. 3. Stone-masons—Indiana. I. Wolin, Jeffrey A.
II. Title.
TN967.S26 1985   553.5′16′09772255   84-43154
ISBN 0-253-18515-7

1 2 3 4 5   89 88 87 86 85

For my parents and, of course, Betsy
　　　　—J. A. W.

In memory of my father,
Greeley Ray Sanders,
who taught me the use of hands
and the love of earth.
　　　　—S. R. S.

Grand Central Station, New York, 1984

. . . when I try to imagine a faultless love
   Or the life to come, what I hear is the murmur
Of underground streams, what I see is a limestone landscape.
                              —W. H. Auden

CONTENTS

	Acknowledgments	*xi*

Part One:
IN LIMESTONE COUNTRY
by Scott R. Sanders

1.	Hunting for What Endures	3
2.	Bones and Shells	8
3.	Digging	22
4.	Doorways into the Depths	33
5.	A Veteran	50
6.	Poison	58
7.	The Men in the Trenches	66
8.	Cutting	78
9.	Three Carvers	93
10.	Truth on the Back Roads	107
11.	Stone Towns and the Country Between	122
12.	The Shape of Things to Come	137
	Epilogue: In Praise of Limestone	150

Part Two:
THE STONE BELT
by Jeffrey A. Wolin

1.	The Land and Its Transformation	155
2.	Quarrying	177
3.	Stone Mills	199
4.	Works	227
	Afterword	254

ACKNOWLEDGMENTS

Jeffrey Wolin wishes to thank the many people who encouraged and supported him in this work. Linda Reilly first published some of the images from this book in *Light Impressions Review* (September 1982). Linda McCausland was the first person to show a selection of these images at her gallery, "28 Arlington." Marianne Fulton obtained some of these images for the George Eastman House. Owen Butler of Rochester Institute of Technology, Reg Heron, Kent Dawalt, and other colleagues in the Fine Arts Department at Indiana University gave thoughtful advice and criticism at key times. Mort Lowengrub and the Office of Research and Graduate Development at Indiana University made it possible for Wolin to enlarge the scope of this work to include looks at uses of stone outside the stone belt.

Scott Sanders wishes to thank the National Endowment for the Arts and the Indiana Arts Commission (and the long-suffering taxpayers who fund both programs) for supporting him with fellowships during the making of this book. He is grateful to Robley Wilson, Jr., editor of *The North American Review*, Erin McGraw, editor of *Indiana Review*, Michael Martone, editor of *Poet & Critic*, and John Witte, editor of *Northwest Review*, for permission to reprint portions of the text that originally appeared in those magazines; to Donald Gray, Paul Strohm, and Mary Burgan, who helped him clear space in his life for writing; and to Ruth, Eva, and Jesse, who put up with dusty boots and a lifetime's worth of talk about limestone.

Many people in the stone belt were especially generous with their time and knowledge. We wish, in particular, to thank William McDonald of the Indiana Limestone Institute, John B. Patton, State Geologist, Donald D. Carr of the Indiana Geological Survey, Jim Kingsley of Indiana Limestone Company, Robert Edinger of Edinger Stone Company, Jim Fluck and Roy Torbit of Fluck Cut Stone Company, Clayton Holmes of Independent Limestone Company, Ted Reed of Reed Quarries, Ed Bennett of Bennett Stone Company, and Claude Black, formerly of Indiana Limestone Company. We thank F. G. Summitt of Summitt & Evans for sharing his considerable knowledge, indoors and outdoors, Wilbur and Dan Bybee of Bybee Stone Company, Pat Fell of B. G. Hoadley Stone Company, and Jack Rogers of Woolery Stone Company, for opening their quarries and mills to this pair of curious observers. These good people will not agree with

ACKNOWLEDGMENTS

everything we say and show in this book, but we hope they will sense the attitude of respect that informs all of our seeing and saying.

We are most deeply indebted to the quarry and mill workers of the stone belt, those named in these pages and those unnamed, who taught us with patience and dignity about themselves and their country.

Part One: IN LIMESTONE COUNTRY

1 Hunting for What Endures

Solid as rock, we say. Build your foundations upon stone, we say. But of course the rocks are not fixed. Waters carve them, winds abrade them, heat and cold fracture them, the twitches of the heaving earth buckle and warp them. The sands on our beaches started out as bits of mountain. The soil that feeds us is laced through and through with the scourings of stones. Right this minute the oceans are manufacturing the stuff, and so are volcanoes. A cauldron of fresh brew is boiling underneath our feet. The very continents glide about like great rafts, floating on the planet's molten mantle, one plate grinding against another, new rocks surging up from trenches in the sea bed, old rocks slithering down. A time-lapse film of any landscape, with frames shot every thousand years or so, would reveal a swarm of changes. From one millennial blink to the next, God would see an altered world. The Psalmist knew what he was talking about when he said the hills skip like lambs. They do, only we're too quick-eyed to notice.

STONE COUNTRY

Still, in our hasty sight the rocks seem fixed. By comparison with our brief lives and our fleeting works, they might as well be eternal. Their clocks are running, but only a millionth as fast as ours. There is nothing like geology to take the urgency out of the morning's news. If we could watch events from the rocks' point of view, all of human history, from the stalking of woolly mammoths to the launching of space shuttles, would appear like a blinding flash. Our longest running shows, such as Egypt or China, would be mere buzzings in the ears of stones. Babylon, Rome, New York: snap, crackle, pop. When we disappear, we probably will carry a good many other animals and plants down to extinction with us. But rocks won't keep much record of our brief transit. For all our drilling and blasting, we have barely scratched this stony planet.

If you find comfort in that, if the rush and sizzle of life makes you hunger for durable goods, if your blood pressure goes up and down with the stock market or baseball scores or political polls, if you fret about pulling off the monthly balancing act in your checkbook, if the world has spun you dizzy—you would do well to spend some time in stony country. And no matter how far you roam, you will find few places where the presence of stone is richer than it is in a narrow belt of hills and creek beds in southern Indiana.

The territory I have in mind lies chiefly in the counties of Lawrence and Monroe. The northern and southern boundaries are marked by the main branch and the East Fork of the White River—a meandering stream that may have been white when the Miamis named it, but that's muddy brown now. The lesser creeks of the region wear names like Jack's Defeat, Beanblossom, McCormick's, Pleasant Run, Leatherwood, Goose, Salt (which is sweet), Clear (which is cloudy), and Big (which is little). The principal towns are the two county seats, Bloomington and Bedford. The villages and wide places in the road include Romona, Spencer, Stinesville, Ellettsville, Clear Creek, Guthrie, Harrodsburg, Peerless, Oolitic, Needmore. On the western edge of the country I'm talking about is Popcorn, on the south are Pinhook and Buddha, just over the eastern border are Nashville and Gnaw Bone, and up near the northern edge are Cuba and Carp.

In this neighborhood the reigning rock is limestone, one of the commonest on earth and the one that wears the shapes of time most handsomely. You enter the limestone country from the east by sliding down the Appalachian Mountains, crossing all of Ohio and half of Indiana, and then nosing downstate on the smaller roads. Coming from the south, you wind through Kentucky, bridge the Ohio near Louisville, and keep on driving through oak and maple forest until you reach the East Fork of the White River in Lawrence County. You enter from the west over the prairies of Illinois and the surly Wabash and the scarred terrain of the Indiana coal fields. From the north, all the way from Chicago, say, or a trailer on Lake Michigan, you roll across a plain, flat as a parking lot, then below Indianapolis, keeping to the valley of the White River, you suddenly reach the rumpled hills that mark the edge of the last glacier, and there in the roadcuts and creek bluffs and lumpy fields you see the limestone showing its silvery gray flanks.

The smoothest and thickest flanks of all belong to the limestone that geologists call Salem. It was laid down roughly three hundred and thirty million years ago, in the Mississippian period, around the time the sharks were getting their start, a little before the cockroaches, long before the appearance of anything vaguely resembling a mammal. The Salem outcrop, which extends northward in a snaky belt never more than fifteen miles wide from the bluffs of the Ohio almost a third of the way up the state, is the largest accessible deposit of premium building stone in the United States, and one of the largest in the world. If you live anywhere in the lower forty-eight states, you are probably within walking distance of a library, bank, factory, church, house, or skyscraper built with Salem limestone. For more than a hundred years now, chunks of southern Indiana have been shipped all over the continent.

Within a few miles of where I sit in Bloomington, there are gaping holes in the earth from which the stone was dug for the Empire State Building, the Pentagon, Rockefeller Center, the National Cathedral, Grand Central Station, San Francisco's City Hall, Chicago's Tribune Tower, the Dallas Museum of Fine Arts, New York's Metropolitan Museum of Art, the Free Library of Philadelphia, Vanderbilt mansions, fourteen state capitols, and countless other buildings grand or humble. Walk to your town square or to the lawn of the nearest courthouse, and chances are you'll find a war memorial carved from Indiana limestone. In Washington, for example, Abraham Lincoln's statue is surrounded by walls of it, and his weighty words are carved into it. Another limestone memorial you've surely seen, at least in pictures, is the statue of marines hoisting the stars and stripes over Iwo Jima. The Bureau of Internal Revenue toils through its endless piles of forms behind sturdy walls of Salem stone, and so do the Departments of State and Commerce and the U.S. Postal Service, along with many other bureaucracies too numerous and disheartening to mention. Right now, the national Capitol—built of crumbly Virginia sandstone—is being given a facelift with the palest and finest of Salem. Over the past century, the destinations for this stone read like a graph of America's growth: first the great international cities, Chicago and Boston and New York; then the muscle cities of the Midwest, Pittsburgh and Cleveland, St. Louis and Indianapolis; now the glittering cities of the sunbelt. New towers, sheathed with Indiana limestone, are rising today in Miami, Atlanta, New Orleans, Houston, Dallas, Denver. There was a time, back in the heyday of the industry before the Depression, when two-thirds of all the cut stone in America was coming from this little strip of land, an area so small you can hike it from side to side or bicycle it from end to end in a single day.

    Digging up that much stone has left a good many holes hereabouts. The work of cutting and carving it has occupied thousands of local men. Everywhere you look, you see the brawny machinery these men used to handle the stone, the humpbacked mills where they shaped it, the rails where they shipped it, the piles where they heaped what they could not use. For the past year, in company with Jeffrey Wolin, I have been talking with the stone men, exploring the mills and quarries, tramping along the creeks and railroad spurs. Each of us had poked around the region for several years before our meeting, but we did our best work, our most intense work, together. This book is our personal map of the limestone country. What Jeff came to understand about the place and its people is recorded in photographs. My own understanding is caught here in these words. We learned a great deal from one another, but in the end each of us offers his own vision. We did not set out, as scholars might, to give a sober history, a skein of dates and dollar signs. Nor did we come to visit this place as outsiders might, searching for the exotic, squinting our eyes for a glimpse of local color. We live here. This is a travel book of sorts, yet it concerns travels undertaken in our own neighborhood. It is also a sort of biography, not about a person—although it is filled with portraits of individuals—but about a place. It is a close look at a piece of the earth where the accidents of geology have bared a special kind of stone, and where landscape, towns, and the people themselves bear the mark of that stone.

2 Bones and Shells

Near the village of Needmore, toward the southern end of the limestone belt, there is a graveyard on a knobby hill surrounded on three sides by abandoned quarries. You reach the cemetery on the fourth side, along a gravel drive. Raspberries, just coming ripe, lapped against the side of our car as Jeff and I rolled up to the locked gate one morning late in June. We had the place to ourselves. From the borders of the cemetery, oxeye daisies gazed darkly at us. Daylilies burned a fierce orange. Yucca plants lifted their pale flickering blossoms.

We had come here in search of limestone gravemarkers, especially ones that might carry the imprint of private feelings. And we were not disappointed. The first headstone we spied was a homemade job carved with unsure hand on a rough slab, for a boy who had died at the age of nine in the last year of World War II. In the upper right-hand corner was the inscription

        ASLEEP
        IN JESUS,

with the J written backwards, the way my six-year-old son used to write it a couple of years ago. In the upper left was a kindergarten sun, light beams radiating out from it in squiggly lines. The picture might have been carved on the wall of a cave fifty thousand years ago, it was so elemental a sign of hope.

    Jeff was on his belly in the grass, taking a photograph of this marker, when a car gritted down the drive and pulled to a stop behind ours. The man who emerged from the cloud of gravel dust was about seventy, a slow-stepper, wearing baggy blue workpants, a white shirt, a baseball cap emblazoned with the Goodyear insignia, and sunglasses that turned his eyes into blank windows. His cheek bulged with chewing tobacco and his right front pocket bulged with a revolver. I could see the blond handle, cross-hatched for easy gripping. Was our visitor trying to drum up more business for the graveyard? As he approached, I kept my eye on his right hand.

    We exchanged howdies, mine cautious, his cold. He leaned to spit, shifted his chaw, shifted his feet, then just stood there behind the blank windows of his sunglasses, as if waiting to be told what in the devil we two fellows were doing with a notebook and a camera in Hopkins Cemetery at eight o'clock in the morning. So I told him we were after limestone.

    As soon as that word hit the air, he thawed out. "They ain't nothing much around here *but* limestone." He shoved hands in pockets—for a quick draw?—and then set off talking. He was the caretaker for this place, kept it mowed, kept the plastic flowers in their holders, kept his eye on visitors. The cemetery plot had been donated to the township by a guy named Hopkins, who sold all the surrounding land to the limestone companies. Quarry holes yawned out there in any direction you cared to look. "There's good stone right under where we're standing. We only dig the graves four feet deep. If we had to go six, we couldn't bury anyone without drills and dynamite." Most all the dead belonged to stone families, whose menfolk worked in the quarries and mills. For markers they favored limestone. "Worked all their lives in it, and now they're buried on top of it and underneath it. Just like a sandwich."

The combination of gun and hand in his pocket still worried me. When he paused to spit, I asked him why he packed it along. "On account of that Leggo boy who's been hiding around this cemetery since he broke out of jail," he answered. "I don't expect he'd shoot me. I been knowing him all his life. But you can't never tell about kids." Neither Jeff nor I had heard of the infamous Leggo boy, so the caretaker filled us in on the history. He was a local tough, eighteen or so, grew up right here in Needmore. Not long ago he broke into an aluminum stamping plant down the road, a brick leviathan of a building that used to be the Furst-Kerber limestone mill. The sheriff caught up with him, put the boy in the Bedford jail. Next visiting hours, Leggo's mother came to see him with a pistol in her purse. The sheriff found the gun, got mad, sent the boy to the state penitentiary for small-timers, up near Indianapolis. Next visiting hours the mother came to see him, and blame if Leggo didn't walk right straight out the door with her, pretending he was just a visitor.

Since the jailbreak, the mother had been making frequent trips down the gravel drive of the cemetery—"parks about right where your car's sitting"—to leave bags of groceries for young Leggo. "I seen him in the woods and the quarries three or four times, but never close enough to grab. The police had squad cars and a helicopter out here looking for him the day before yesterday, but they never did find him." Pause for a spit. "He could hide in them quarries for ten years and nobody'd ever run him out. There's old hermits and guys have *lived* out there, winter and summer, wild as bears." The caretaker was coming down to check on things at the cemetery every hour or so, to keep an eye peeled for the kid and the old lady.

"She give him a gun, of course. But I don't think he'd shoot anybody except the sheriff. The sheriff, now, Leggo'll shoot *him* on sight. You boys should be all right. Just so long as he don't take you two for a couple of detectives, you don't have to sweat."

When the caretaker was gone and Jeff was back at work with his camera, I sweated anyway, less on account of June than on account of Leggo. "Keep me covered," Jeff said. We discussed which of the pair sounded like the meaner customer, the jailbird or the mother. I felt like ringing a bell and declaring that we were not detectives, that we were neutral in this whole affair. I kept listening for the grit of Mrs. Leggo's tires on the cemetery drive, for the click of Leggo's gun. Instead, from across the field of quarries, I heard the humph of the aluminum-stamping plant, humph humph, like an asthmatic dragon.

I browsed among the graves. The ground felt spongy under my boots. Excuse me, old souls, for treading on your roof. Limestone markers from last century lay tumbled, heaved over more likely by frost than by vandals. The names were mostly ones that people around here still wear: Grayson, Patton, Campbell, Turpin, Swango, Holtsclaw, Sears. Joseph Massey, "Nipped in the opening bloom of youth," gone now almost a century. Roy Black, "BOR Feb 23 1911 DID Mar 3 1913." On the headstone for a child, dead at four, rested a small wooden car. Rain and sun had bleached the wood to the color of ivory. I lifted the car, amazed to find it loose, amazed that no one had stolen it in the years since the child's death. I set it down again, scooted it back and forth to make sure the stiff wheels still rolled.

STONE COUNTRY

All the while we prowled through the cemetery, I imagined Leggo out there hiding in the crevices of stone. There were plenty of crevices to choose from. The quarries to the north were shallow and scrubby, grown up in sycamores and sumac. Immediately to the south opened the vast gulf of the Empire State Building hole. I stood on the lip staring down a hundred feet or so into the green water, and tried to stuff the skyscraper, block by chunky imagined block, into that enormous pit. Moving it back across country from 34th Street in Manhattan to Needmore in Indiana was easy, but seeing how that tower could squeeze into this huge trench was hard. Think of a domed stadium, squash it into the shape of a box, and you will have some notion of the size of the emptiness left by the Empire State Building. Beyond that gulf the gouges and rubble heaps stretched southward for miles toward Oolitic and Bedford, the largest cluster of building-stone quarries in the world. The snort of cranes and the machine-gun rattle of air drills sounded from a working pit down that way.

East of the burying ground there was another sheer drop of ninety feet or so, down to the water of a junk-filled quarry. One glimpse, and I understood the sign we had seen on a hogwire fence when driving up: "Please do not throw trash in the cemetery. There is a dump at the back." Junkyard behind graveyard. Here it was, in the hole left from the building of a bank or a museum. Down below, trash spread across the pit like a river delta—car bodies rusted the color of a rooster's crown, bullet-riddled stoves and washers, televisions, buggies, bottles, a jigsaw puzzle of torn plastic. I remembered one of the stories about how Needmore got its name. (There's another settlement with the same name twenty miles northeast, in Brown County, with the same story attached to it.) When the village was still small enough to shout across, a slick visitor from the East was asked what the place needed in order to become a city, a seat of culture. "I never saw a place that needed more," he answered. Needmore still lacked a lot, if you were looking for a city, but at least it was accumulating the debris of civilization.

Jeff balanced his tripod and aimed his bulky box of a camera dizzily down at the dump. The height made my legs feel mushy. The ledge we were standing on was actually a wall of old quarry blocks, erected here twenty years ago to keep the easternmost graves from slipping over the brink. Some poor souls took the dive before the wall was built. I thought about the old quarriers, working down there in the hole and glancing up one day to see a rain of bones.

The distance from where we stood on the brink of Hopkins Cemetery down to the scummy water of the township dump is about as thick as the Salem deposit ever gets. Ninety feet of rock, the residue from several million years of living and dying. Like most limestone, this local stuff is a cake of corpses, a hardened graveyard of sea creatures. When it was formed, the area we now call the Midwest was covered by a tropical ocean, very much like the seas today around the Bahamas, where brand-new limestone is forming. Three hundred and thirty million years ago, southern Indiana was southern indeed— several hundred miles below the equator, which ran diagonally across the land mass that would become North America, from what is now southern California to what is now Lake Superior. On land in those days, flitting among ferns and horsetails, there were the ancestors of moths, spiders, crabs, flies. Reptiles, the first beasts with backbones that could live entirely out of the water, had not yet appeared. They would arrive later, with the swamps that gave us our coal, and after them would come the dinosaurs. Along the shore amphibians were slithering in and out of the water, undecided where they belonged.

STONE COUNTRY

But Indiana was several hundred miles from the nearest shore. The inland sea stretched from what is now Nebraska to what is now Pennsylvania. If you had looked east from the Pennsylvania beach, you would have seen snow-capped mountains. To the west, you would have seen nothing but water. The sea was shallow and warm, rich in calcium, and it bloomed with animals that had shells instead of backbones. As these creatures died, their husks settled to the bottom, where tidal currents and wave action wore them to bits and sorted the bits according to size. This limey sand was gradually cemented together by calcium carbonate, layer upon layer, like a giant wedding cake.

If you count the deposits above and below the Salem, the limestone in this region runs more than four hundred feet deep. The Salem is special because it is fine textured, strong, easily worked, without any apparent grain, and above all because it was laid down in great thicknesses. You can dig up fifty-ton blocks of it that are as uniform in texture and color as bread. The overlying and underlying deposits were flawed by shifting conditions, the ocean sometimes too deep, sometimes too shallow, mud oozing in, clumpy coarse shells jumbled together with fine ones. Salem limestone owes its virtues to the fact that it was deposited in dull times. For millions of years, nothing much changed. On both coasts mountain ranges were rising and falling, volcanoes were puffing, the basement of the continent was buckling and faulting; but here in the inland sea all was serene. Waves, tides, sunlight, trillions of sea-beasties—and a ninety-foot layer of lovely stone.

There is nothing inherently placid about limestone. The cap of Mount Everest is made of it, heaved miles into the air when India plowed into Asia. The difference is that the Himalayas are near the lip of a continental plate, while the hills of Indiana are in the middle of North America. But sooner or later change comes even to the drowsy interior. Some two hundred and fifty million years ago, the earth-shrug that heaved up the Appalachian Mountains elevated these limestone beds above sea level, tilting them so that progressively older layers were exposed to the weather and to curious ramblers. If you hike from west to east across the unglaciated ridges and valleys of southern Indiana, you will traverse belt after belt of limestone, ranging in age from about three hundred to about five hundred million years. There are places where you can leap from one outcropping to another and cross a gulf of a thousand centuries. Think of a deck of cards, with individual cards representing beds of sedimentary rock, the oldest at the bottom, newest at the top, then tilt that deck ever so slightly to the south and west, then let wind and rain scour the bared edges for a couple of hundred million years. The Indiana limestone country is the narrow belt where the edge of the Salem deposit, after all that erosion, is raggedly exposed. Geological Survey maps show the area of the outcrop to be roughly 28,000 acres. Of these, some 25,000 are in Lawrence and Monroe Counties: if you want to quarry this stone in daylight, here is where you have to dig. If you cared to go underground, you could follow this bed all the way to the center of the Illinois Basin, two hundred miles west of here. With the stone you found below, you could rebuild all the cities of earth.

I have checked this general picture of things with the State Geologist, John Patton, who ought to know. He's responsible for thinking about all of Indiana's mineral resources, but limestone is obviously his passion. Ask him about it, and you might have asked a coon hunter about hound dogs or a gourmet about food. He talks with a young man's zest, and appears a decade younger than his real age, which is nearly seventy. He looks remarkably like William Faulkner in the years after the Nobel Prize, a dapper, fit, poised man with silver hair swept to one side and a silver moustache. You can see his mind working between sentences, checking every word to make sure it's plumb and true. Like Faulkner, he smokes a pipe, burning tobacco he raises himself. In the office he is a dandy, wearing black wing-tip shoes, an immaculate shirt, a charcoal suit with a red kerchief flowering from the pocket. In the field he wears spotless khaki trousers, blue workshirt, and a Panama hat. With a lump of sandstone he scours the rust from his rock hammer. In a plastic vial he carries sulphur for discouraging chiggers. Only his outdoor boots give him away. They are the scuffed, heavy brogans of a man who likes to tramp.

That love of fieldwork led him to become a geologist. In northern Indiana when he was growing up, few people had even heard the word geology. But a high-school friend had a father who explored for oil, so Patton got an inkling of what such work might be like. At Indiana University, majoring in chemistry, he took as much geology as he could fit in. When he graduated, Du Pont offered him a job in their labs as a research chemist. Could he also work outdoors, look for minerals, he asked? No, they said. They wanted him to wear a white coat and stay at his bench. This was in the deepest trough of the Depression. Even bad jobs were scarce, and here was a prize one. Broke, uncertain, Patton searched his soul for a while, then finally turned Du Pont down. He would not spend the rest of his life indoors. By hook and crook he went on to finish a master's degree, this time in geology, and landed a job with Standard Oil, prospecting on the Gulf Coast. Derricks, miles of drilling pipe, the hunt for oil: memories of that work fill him with pride and pleasure still. After the Second World War he was lured back home to Indiana, to set up an industrial minerals section for the Geological Survey. Like the oil in his wells, he rose steadily, becoming the State Geologist in 1959.

In his passion for limestone John Patton follows in a venerable tradition. The very first State Geologist was David Dale Owen, son of Robert Owen, the visionary Scottish industrialist who bought the Indiana town of Harmonie in 1825. The elder Owen renamed the town New Harmony and set out to make of it a utopian commonwealth. Into the backwoods of Indiana he shipped a boatload of scientists and educators, among them another Scotsman, William McClure, who turned his New Harmony house into a geological museum. After staring at the specimens in McClure's parlor, David Dale Owen gave up the study of art to become a geologist. In 1838, reporting on *A Geological Reconnaissance of the State of Indiana*, Owen the younger noted the presence of the limestone and its fitness for building. He lumped together some thirty million years' worth of formations under the single label of Mountain Limestone, but he was pointing in the right direction.

"You've got it more or less correct," Patton told me about my picture of the way Salem limestone was made. "We can't be sure about the details. Geology is a science of inference. It's detective work, and that's what makes it challenging. Putting together an account of the geological past is like describing a shipwreck when all you have to go by is the debris that is washed ashore. You will find spars, corks, bits of luggage; but the ship itself has gone down."

On a clear, blistering July morning, he took Jeff and me out to see some of the wreckage, the scraps of geological evidence, in roadcuts and quarries near Bedford. The highway was ribboned blue with chicory. Even where the stone did not show, Patton traced the underground shape of it with gesture and voice. The soil over the bedrock here is thin. Trees do well enough on such puny fare. Much of the

STONE COUNTRY

limestone country is furred over with forest, splendid tulip trees and oaks, maples and dogwood and beech. Trucks bearing fat logs to market ride the roads. Where the trees have been cleared and the hillsides plowed for corn, the topsoil, never more than a few inches thick, soon washes away, leaving a red clay that turns to clinging goop in wet weather and to ruddy concrete in dry. This clay, called *terra rossa*, is the residue of minerals left behind when limestone dissolves. Because limestone is soluble, the rain weathers it into a characteristic landscape of sinkholes, caves, springs, and underground rivers. The sinkholes are especially striking, steep bowls in the ground that might be a few feet across, or a few hundred. Some are ponds, unblinking green eyes, thick with frogsong and the dry-leaf flicker of butterflies' brushing wingtips. Some are open at the bottom, like unplugged bathtubs, and everything in the vicinity—trees and dirt and cows—slides down into the earth's raw throat. In parts of the limestone country, there are more than a thousand sinkholes per square mile, pockmarks thicker than craters on the moon. Deep winter is a good time to see this topography, when landforms are clarified and simplified by snow. But even in high summer, when Patton showed us around, the shape of water-carved stone is visible through the skin of pasture and woods, like a skull in an aged person's face.

In a few places you can see what the bedrock would look like without its thin flesh. One of these sites is a ledge beside an old quarry in Peerless, not far south of the cemetery where Leggo was hiding. Patton took us there as well. This ledge had been stripped of its covering of soil with high-pressure hoses back in the 1920s, but the operation was abandoned before quarrying began. For the next forty years, what you would have seen there was the earth's stony crust, eroded and cracked, a maze of upthrust hummocks and twisting runnels, like the gray convoluted surface of a brain. The washed-out crannies between hummocks are called grikes. The upthrust knobs are called lapies. "Don't they look like the heads of elephants?" said Patton. They did, all the more so because they had recently been engulfed by the green of the returning forest. In the last ten years the ledge had caught some dirt and now sprouted a young thicket of redbuds, dogwoods, sassafras, and wildflowers. On the July morning of our visit the butterfly weed spread its brilliant orange starburst of flowers, and lived up to its name by attracting squadrons of swallowtails.

Coming away from the ledge at Peerless we met an old man who leaned on a walking stick and carried a bulging gray trash bag. His face, sagging inward like a sinkhole over his bare gums, was covered with a sheen of sweat. "Yellowroot," he told us, opening the bag to display a snarl of bright yellow roots as thin as whiskers. "Folks also call it goldenseal. It's a medicine." Did he use it? "Not me. Never. But there's a store over to Paoli that buys all of it I can dig." He trudged off, bent over his staff, and we snaked our way back to the highway.

On roadcuts and outcrops, Patton identified the Salem bed from a distance by its massiveness, its clear lines, its habit of weathering in rounded, strokable shapes. Up close, banging a chunk loose with his hammer, he showed me the fine grain of it through a magnifying eyepiece. I felt like a jeweler, gazing at a tissue of diamonds. "What you see glittering there is calcite, the crystalline form of calcium carbonate. If you look closely you can see fossils of the typical fauna." He rattled off names: brachiopods, gastropods, pelecypods, and other pods; frail bryozoans, trellises of pure chalk; foraminifera, like spiral galaxies; crinoids, miniature relatives of the starfish, five-armed sea lilies with the segments of their stem shaped like washers for a quarter-inch bolt; and, most common of all, the tiny *Endothyra baileyi*, less than a millimeter in length, looking under the glass like doughy pretzels or chunky sponges.

The geological view of things is not universally shared in southern Indiana. One day, sitting on a limestone wall across from Summitt's Grocery in Stinesville, I fell to talking with a refrigeration man named Bill. About forty-five, built like a pole-vaulter, with the handsome engraved face of a television cowboy, he used to work in the quarries, but he had given that up twenty years ago for the steadier work of keeping things cool. "What always stumped me was how these bugs got in this stone," he said, pointing at the lacy web of a bryozoan on the wall. "You'd be lifting out a block from way down under ground, and there you'd see all these curly bugs. Where'd they come from?" I brought up the subject of fossils, and got a blank stare. I threw out the figure of three hundred million years, and drew an even blanker stare. Then—unable to let a question go by if I think I have the answer—I spoke about continental drift, glacial advance and retreat, the dancing of the equator, and ancient oceans. The mention of water finally roused a smile of recognition from Bill. "The Flood," he said. "You're talking about Noah's time. So that's when God put the bugs in there?"

Even the men who operate quarries do not always believe in geology, John Patton told me. "Some of them, when they're considering where to make a new opening, use geology like a rabbit's foot. They suspect there might be something to it, and so they come and talk with us, for good luck, then they go right ahead and drill where they had been meaning to drill anyway." The results of these exploratory drillings are cylinders of rock, as thick as baseball bats, which break into chunks the length of your foot or your arm. When Patton and his colleagues are called out to examine these cores, the sections

Bones and Shells

might be strewn topsy-turvy all over a field. "After the dogs and the kids have been rolling the cores, nobody can tell you which end is up or which is down." One quarrier would often drill in formations that underlie the Salem, then complain when he found no good stone. He absolutely refused to believe that sedimentary rock was bedded down in layers. He thought the creamy deposits of Salem might turn up anywhere, like buried treasure. You just had to sniff around for it.

With or without the help of geology, men have been finding good stone in these hills and creekbeds for a hundred and fifty years. Jeff and I went sniffing around in their tracks, looking for the holes they left behind.

3 Digging

The earliest quarries were not dug into the ground, but hacked out of ledges. Before about 1880 there were no machines for cutting down into the rock, so quarriers pounded holes in bluffs and outcrops with sledge and drill, tamped in black powder, blew a chunk loose, sliced it into blocks with a crosscut saw, and then chipped the blocks as square as they could with short-winged picks called scabbling hammers. Grunting men hoisted the stone onto wagons with ropes and pulleys. Ox teams yoked to wagons with twelve-foot wheels hauled it away to farm or town, where it was used for chimneys, bridge abutments, monument bases, foundations, flagging. You can still find some of the scabbled blocks in the footings of bridges along White River, the faces of the stone looking as if they had been scratched by steel-clawed tigers. But after weathering for a century, the marks where the digging was done have been nearly erased.

In spring, when mayapples were unfurling and fiddlehead ferns were breaking the surface like sea serpents, Jeff and I hiked along Jack's Defeat Creek to look for the site of the oldest quarry on record, a ledge worked by Richard Gilbert in 1827. Of course, we had asked around about the naming of the creek. Most of the stories featured one Jack Storm, some versions claiming he had got his horse mired in the mud, some that he had lost a clutch of cows when trying to ford them across the stream in floodtime, some saying that Jack himself had swum over to get drunk and on the return trip had drowned. One way or another, Jack had been defeated, and here was the name reminding us of the fact forever.

The spring had been a wet one, and the creek was licking its lips. The soggy bottomland flared yellow with celandine poppies, white with rue anemone and bloodroot, pink with spring beauties, purple with trillium, blue with phlox. It was mushroom weather. The only souls we saw in the woods were an older couple, willowy and silent, working their way up a hillside through the mayapples, the woman turning over leaves with a forked stick, the man shuffling along, green shopping bag hanging from his belt on one side, a holstered gun on the other. I waved, but they never saw me. They were intent on finding morels, the creamy, wrinkled, upthrust thumbs of spring.

We were intent on finding Gilbert's quarry. We knew from studying topographic maps and old books where it ought to be. When we came to the place, a stairstep of ledges tufty with moss and foamy with waterfalls, we could at first see no sign of quarriers. There were abundant signs of more recent visitors. Cloudy plastic from a makeshift tent, fire rings, cans and cartons enough for a week's camping. The stone cantilevered out in slabs two feet thick, begging to be plucked. "If Gilbert didn't dig here," said Jeff, "he missed a good bet."

STONE COUNTRY

On hands and knees, awkward goats, we clambered over the ledges, hunting for drill marks. Nothing, nothing. Only cushiony moss, ferns still wrinkled like newborns, weathered stone. The waterfalls consoled us. We were on the point of leaving when Jeff checked one last time along the creek at the base of the ledge. He gave a shout, and I came running. There in the shallows, its gray topknot breaking the current, was a tumbled block with the rounded slots of drill holes down one edge. When Gilbert's men put those marks there, the Shawnees had only just surrendered this land to whites, John Quincy Adams was president, and bears still lumbered through the woods of Indiana. So here was the beginning, the first harvesting of stone we knew about.

From Gilbert's time until after the Civil War, not much changed in the methods of digging stone. Drill, blast, saw, and scabble. Hauling it became easier after the railroads began spidering into the region in the 1850s and 1860s. The oxen only had to pull the wagons to the nearest track and turn over their burden to locomotives. The New Albany and Salem line began carrying traffic along Jack's Defeat Creek in 1854, delivering stone as far away as Louisville. Quarries opened up all over the region, by grace of and for the sake of the railways, supplying stone for bridges and roadbeds.

A few miles upstream from Gilbert's ledge, we found one of these old railroad quarries. Here the human marks were unmistakable. Cut back into a bluff, the mossy face of the wall looked like a billiard table turned on its side. To explore it, we had to interrupt the lawn-mowing of the landowner, Jim Medley, a lanky slow-talker in his late twenties. Judging by the trim looks of his place—tractor shed, new orchard, sprouting garden, bandbox house—he did not interrupt his work very often. He crawled off his mower, wiped his hands on a rag, and apologized for the shape of things. Things looked good to me, but what he saw around him was still only a glimmering of the grand estate he carried in his mind. The house, for example, was only a shell around a trailer. A few years earlier, wanting to make a home for his wife and kids, he had started with the trailer. Then he put a roof over it, then a porch beside it, then two rooms in front of it, then walls around three sides of it. The fourth side he had left open, because one day, when the money was right, he was going to haul the trailer out and leave behind nothing but the house, a proper house. Money wasn't right just then, since he earned his living as a carpenter, and he'd been laid up all winter after breaking his elbow playing church basketball on the concrete floor of the Salvation Army gym.

"I'm planning to put a pole barn in that old quarry as soon as I can get it filled up and them stumps out." He pointed at a pair of two-foot-thick stumps, one from an ash, the other from a sycamore. When the quarriers abandoned this pit, the bottom would have been raw stone. And now here were the fat rumps of huge trees. You could see where Medley had been clawing at the roots with a backhoe. Studying his lot, you could also see why he needed a pole barn. Keeping the backhoe company were a bulldozer, two tractors, a pickup, a dump truck, two cars, a grader, a log wagon, and the lawn mower. He was equipped to build a whole town all by himself. "I hate covering up that green wall. I like looking at it, and all the flowers. But it's the only place I can put my barn."

The moss on the wall was so deep piled I could plunge my thumb in up to the socket. The floor was so thick with blooms—twinflowers, poppies, anemones, violets, Dutchman's breeches, squirrel corn—that I felt guilty for treading there. A log fallen slantwise into the pit and rotted almost to loam was a highway for possums, Medley told us. "They come along there morning and evening, regular as buses." Overhead, in the walnuts and shagbark hickories rooted along the quarry's lip, red-bellied woodpeckers and white-breasted nuthatches were grazing for bugs, every half minute or so reminding one another with shrill whistles of their species and their whereabouts.

"I sure do hate covering up that old quarry," Medley repeated when we got ready to go. "In the wintertime you can look out there at that green wall, where the snow don't stick, and it's like seeing a hunk of spring."

Beginning in the 1870s, many of the blocks rolling by on flatcars past where Medley would build his bandbox of a house were destined for Chicago. The heart of the old wooden city had burned down in 1871, and now a new city of stone was rising from the ashes. The Indiana quarriers should have included Mrs. O'Leary and her clumsy cow in their prayers, because the tipping over of that lantern in Chicago was a godsend for the industry. Trainload after trainload, the new Chicago streamed out of the limestone country, blocks for City Hall, for museums, mansions, libraries, churches, hotels, thick blocks at first, since the outer walls in those early days bore all the weight—the groundfloor walls for the Monadnock Building (1891), sixteen stories high, were fifteen feet thick—then after a while thinner blocks, when Louis Sullivan and other Chicago architects learned to put up towers of steel and to use stone and brick no longer for the skeleton but only for the skin, buildings higher than the Tower of Babel, skyscrapers, trainload after trainload of Indiana limestone, streaming not only to Chicago but to New York, Boston, Baltimore, Washington, St. Louis, Cincinnati, Philadelphia, Pittsburgh, Cleveland, Detroit, all the burly northern cities. America had come down with building fever.

The fever would hit again in the 1920s, again after World War II, again in the 1980s. Each of those periods would be healthy for limestone. But the frenzied era of city-building in the last quarter of the nineteenth century was the birthing time for the industry, the time when the silvery gray and tawny buff rock from Indiana became the darling of the nation's architects. Historians have called this boom time the Age of Energy, the Gilded Age, the Great Barbecue. Immigrants were lapping at the shores of America with every tide. (The buildings they poured through, on Ellis Island, were trimmed and ornamented with Indiana limestone.) Robber barons were busy robbing the public and enlarging their empires. Railroads were stitching the nation into a web of steel, and new terminals rose in city after city like temples for the gods of machinery. The Great Lakes became highways for ships carrying wheat and iron ore and coal. Bridges were needed for the trains, breakwaters for ships, post offices for the torrent of mail, courthouses for the wrangles of the law, office buildings for the masterminds of business and tidy minds of government, banks for the money. Stone, stone, and more stone.

To satisfy that hunger, new quarries opened from one end of the belt to the other. Dark Hollow, Balbec, Spider Creek, Hunter Valley, Blue Hole, Buff Ridge, Reed's Station, Tanyard, Brickyard, Empire, Star, Crescent, Eagle, Preacher's—they bore the names of owners, of places, the names of the shapes they left hollowed in the rock.

By this time there were new machines to do some of the work formerly done by muscle. Although today's steels are harder and the motors stronger, although horse power gave way to steam, and steam to electricity, the methods of quarrying have not changed fundamentally since the nineteenth century. First a promising location must be found. This is a tricky job, one-third science and two-thirds hunch. While there is limestone everywhere in this vicinity, it may be covered beneath too much dirt or lie in beds too thin or suffer too badly from water damage to be suitable for quarrying. The stone men look at outcrops and study core drillings; but in the end, like psychoanalysts probing the unconscious, they have to guess at what lies hidden underground. Opening a new quarry has always been a gamble, and one that often fails. Many a man went bankrupt from digging in a bad spot.

Once the place is chosen, the overburden of soil and waste rock must be stripped away. In the early days this was accomplished by pick and shovel, horse-drawn scrapers, and an occasional charge of black powder. The old-timers called the men who did the stripping the cat-owl gang. It was usually winter work, cold and wet. After ten hours in the muck, the cat-owlers staggered out wearing half their weight in red clay. Horse-scrapers were replaced by steam-powered draglines and shovels, shovels by high-pressure water hoses, hoses in turn by frontloaders and bulldozers and dump trucks. For digging down into the grikes—the mud-filled crevasses—no machine has yet been devised to replace a strong back and a shovel. Today, when the explosives are newfangled slurries and plastic concoctions instead of black powder, the waste rock is still loosened with blasting—"shooting off"—but gingerly, so as not to spoil the underlying bed of usable stone.

Once you get down to that good stone, you need a whole new set of machines. The one that made open-pit quarrying possible is a great mobile engine called a channeler, introduced in the 1880s. Powered by steam or electricity, it chugs on rails from one side of the bed to the other, slamming steel chisels down into the stone, cutting deep slots. The early models, named after their inventors—Wardwell, Sanders, Ingersoll, Sullivan—cut down only six feet, but later the depth was increased to ten or twelve. Hammering and puffing along, they look, sound, and smell like midget locomotives. Their racket turned many of the old-timers deaf. Another way to cut the bed is with a Rube Goldberg contraption called a wire saw, which came along in the 1930s. You would have to see one to understand it or believe it, but essentially a wire saw works by dragging a thin steel cable across the stone. A pair of stands, slender and sturdy, held rigid by guy wires, keeps the wire taut and lowers it on pulleys as the cut deepens. The actual cutting is accomplished by sand, which is fed onto the wire in a watery paste that looks like well-cooked hominy grits. The newest gismo, in use today at Eureka and Dark Hollow, is a rail-guided chain saw with a ten-foot blade and diamond-tipped teeth.

By one machine or another, by chain saw or wire saw or channeler, the quarriers eventually slice the bed into a grid of blocks, the way you might slice a pan of brownies with a knife. You know how tough it is to pry out that first brownie without mangling it. With limestone, it's worse. The first piece to be removed is called the keyblock, and it always provokes a higher than usual proportion of curses. There is no way to get to the base of this first block to cut it loose, so it must be pried, splintered, and worried at, until something like a clean hole has been excavated. Men can then climb down and, by drilling holes and driving wedges, split the neighboring block free at its base, undoing in an hour a three-hundred-million-year-old cement job.

Once there is room for maneuvering, the big cuts—some of them weigh 250 tons—are tugged onto their side, the massive fall cushioned by pillows of loose rock, and then split into mill-sized blocks. The men who clamber over the great capsized stone look like whalers carving blubber from Leviathan. The boss of the crew draws lines in red chalk where he wants it split. A second man bores holes along that line, six inches apart, six inches deep, with an air-powered drill, guiding the bit at the start of each hole with the instep of his boot. A

STONE COUNTRY

third man drops a wedge and on either side of it two sleeves—a slip and feathers—into every hole in the row. Then a big-shouldered man comes along with a six- or eight-pound breaker's hammer, which looks like a truncated prospector's pick, and bangs on each wedge in turn, hard at first, then more gently as the pressure on the rock increases. Toward the end, when the stone is ready to yield, the gang leader and breaker get down and peer at the wedges, deciding just where to place the final few licks. Then tap, tap, and suddenly the huge rock splits the way a ripe watermelon will when a knife is thrust in its green side. More than once, at that moment of sundering, I have heard men yell, "Thar she goes!" Curious, they lean over the side to see how straight a crack they have made in their stranded whale.

Men called hookers then chip holes—dogholes—with scabbling hammers into opposite ends of the block. Cranes or derricks lower two hooks, weighty enough to serve as anchors, and these are fitted into the dogholes. Cautious, a man stands on each hook while the slack is drawn out of the cable overhead. When the claws are firmly seated in the dogholes, the hookers leap clear, engines roar, and up the stone rises on its twisted thread. As the shadow passes, everyone stands clear, for sometimes the hooks snap loose.

The flawed stones, not suitable for use at the moment but possibly salvageable in the future, are stacked in piles called grout. The random waste, called spalls, is heaped wherever there's room. Good stone is piled beside the hole for seasoning, and from there trucks haul it to the mill. The man who decides what's waste and what's good is the stonemarker, who looks over a new block the way a horse trader studies the teeth and legs and coat of a stallion, searching for blemishes. These might be subtle changes in color or texture, coarse patches of fossils, hairline joints, or the inky scrawls along bedding planes that geologists call styolites and quarriers call crow's-feet. A new block has to season because limestone is porous, and when it comes out of the ground it is filled with water or quarry sap. This porosity accounts for the variations in color and for the length of the quarry year. Because it's soggy when it surfaces, a freshly dug block will crack and spoil if it freezes before drying out. That's why quarrying doesn't start until around the middle of March, and ends around the first of November. Stone cut from below the water table is silvery gray, like a vintage Rolls Royce. Stone above the water table picks up iron from the rain and snow-melt soaking down through it, and the oxidation of this iron turns these higher ledges a pale buff, the color of ripe wheat. Sometimes gray stone has been in fashion, sometimes buff, sometimes a mixture of the two. Gray was popular around the turn of the century because it didn't show the soot from coal smoke.

And so the quarriers work across the bed, wedging and splitting, hoisting and hauling, until an entire layer of stone has been removed. Work then resumes with the channeler or wire saw, another depth of ten or twelve feet is carved out, another and another, on down level by level until the usable stone has been exhausted. Since the Salem deposit is typically thirty to seventy feet thick, most quarries are between three and seven floors deep. Each floor is marked by a narrow ledge, for on each new level the channeler or wire saw must start its cut a foot or two further in from the existing wall. Chiseled down far enough into the bedrock, these stairstepped walls would draw nearer and nearer to one another and eventually meet. If you greased that abysmal pit, poured it full of plaster, then pulled the plaster out and set it with the point sticking skyward, you would have a titanic pyramid.

You may well have no use for a pyramid. But the ancient Egyptians did, and they built theirs out of limestone. There's a pyramid under construction right now between Needmore and Oolitic, in the heart of the quarry district. If you had a good arm, you could just about hit it with a baseball from the cemetery where Leggo is hiding. Only two tiers high so far, stalled because the promoters ran out of money, this Indiana pyramid is supposed to be a one-fifth scale model of the one at Gizeh. It's part of the Limestone Tourist Center, which also promises a replica of the Great Wall of China. The brochure, written for a swarm of sightseers who never came, sings the praises of Indiana stone: "It's durable, it's beautiful, and it's natural. The material for the ages." Every now and again the national news media pick up this story and have a romp with it. Look at these crazy Hoosiers building a pyramid! But is it as crazy as building plastic burger-joints in the shape of castles and taco-joints in the shape of haciendas along the highways? And is it half as crazy as building a Pentagon in Washington? If the monument is ever completed at Oolitic, nobody will crawl inside it to plan our extermination. Nobody will demand half our taxes to feed it. No protesters will ever have call to wail or weep on its steps. It seems to me entirely reasonable that people who live near a quarry should decide to raise a pyramid. One is the negative image of the other, the stairstepped pile of stone above ground and the stairstepped hole below.

    Digging deep into the earth and stacking high into the air: it is a labor as old as cities. The Egyptians honored it. Their hieroglyph for art is a stone worker's drill. The sober tombs along the Nile and the wacky Hoosier pyramid near Goose Creek and the seven-story holes in limestone country will be there long after this book and its makers and readers have turned to dust.

4 Doorways into the Depths

As we drew up to the brink of the old quarry, bullfrogs flopped into the water, a slimy brew mint-green with duckweed, and a painted turtle cruised into hiding, its shell bright red around the edges as if colored by a child. A vine clinging to a log turned into a blacksnake and slithered away. It was a small hole, about the size of a three-car garage, five floors deep, and rank, fecund, a swarmy pit where life might start all over again on its evolutionary spiral. The season was late May, the ginger and jack-in-the-pulpit newly out. A dogwood growing on the narrow lip between the second and third floors was dropping a blizzard of petal-like bracts on the water. Presently the bullfrogs resumed their croaking. From a limb above the water a pewee whistled its name, over and over. A beggarly scrawking leaked steadily from a woodpecker hole in a dead tree, where a chick with a brown fluffball for a head was calling fiercely for grub. Jeff and I had to raise our voices to make ourselves heard above the racket of spring.

STONE COUNTRY

We were hunting a place that would give us a feel for what the life of a quarry hand might have been like in the long-gone days. We found it here, at the Old Statehouse Quarry on McCormick's Creek. The site is less than half a mile above where the creek empties into the White River. If you come by water and perch on a flat rock in the middle of the current, you can see why they chose to dig in this place. The bank is a solid bluff of limestone forty feet high, slightly undercut by the stream, like a smooth gray paunch. It is so unblemished that even ferns and mosses find few places to curl their toes. The soil on top is thin, easily stripped. On the day I sat there balanced on a teetery rock in the middle of McCormick's Creek, admiring that bluff, a fine mist of pollen swirled in the air. Driven slantwise by the wind, it looked like gossamer filaments in the sunlight, as if the threads of the universe were showing.

Leaving the bluff in place to keep the creek out, the quarriers began their hole about twenty feet in from the bank. They started work in the fall of 1879, digging stone for the new Indiana capitol. (If you snoop around the Indiana Statehouse, you will find a plaque listing all the political bigwigs who held office during the decade of construction; you will find memorials to presidents, generals, pioneers, a temperance leader, a poet, Indians, and coal miners; but you will not find a word anywhere about the men who dug or cut or laid up the stone for this twelve-acre building.) The working day was ten hours long, and the pay was fifteen cents an hour. For boys, who carried buckets of drinking water slung from yokes around their necks and gathered up tools and passed on hand signals, a day's wage was seventy-five cents. Men and boys, they had grown up on farms and took it for granted that by nightfall a body would be too weary even for swatting mosquitoes. Many of them had come to work in the quarries from scratch-dirt places where the earnings were poorer and the hours longer. When it rained, the pit filled up like a tub and they sloshed about in water. On frosty mornings their hands stuck to the tools and the stone. When the sun poured down they baked in their rock oven, cut off from breezes, the walls holding the heat.

At night they slept beside the quarry in cabins raised on piers, the few married men in private huts, the single men in a communal dormitory. The nearest towns were Spencer, by the dirt road, and Gosport, by the river, neither one close enough for walking to after a day's work, neither one very much of a place when you got there. So in the evenings they sat in that malarial bottomland beside the creek, telling stories, drinking, gambling, singing ballads, drinking, playing mouth organs and banjos, and drinking. They often pooled their money to buy whiskey in bulk, ordering it in by the barrel. Whatever money they had left over they loved to bet on cards or dice, fistfights or races. If they had something to get rid of—an extra gun, a watch, a foot-broken pair of boots—they would wait for payday, sell chances, and then raffle it off.

When they finally did get to town on weekends they found some dry-goods stores, a few churches, half a dozen clubs, and plenty of saloons. Often the stores belonged to the men who owned the quarries, so the workers, buying tobacco or socks, paid their wages right back to the bosses. The old-time quarrymen made little use of the churches. They had a reputation for godlessness, and lived up—or down—to it. For a while an eager minister from Bloomington lured them in fair numbers to his church, but he was soon turned out by his congregation, who objected to having all these rough customers in the pews. The quarriers saved their zeal for the fraternities, becoming Odd Fellows, Knights of Pythias, Red Men, Modern Woodsmen. (A limestone plaque on the Improved Order of Red Men Lodge in Spencer shows a peace pipe, a tomahawk, and the profile of an Indian who looks like a cross between a Roman gladiator and an Incan chieftain.) The saloons provided them with fresh settings for the familiar diversions of gambling and drinking and scrapping. Come Sunday night, back they went to their cabins beside the quarry.

Even though elevated on piers a yard or so above the creek bottom, the cabins must have flooded every spring. The spring of our visit, the highwater marks on the sycamores were just above my head. Piebald sycamores, liking wet feet, were the only trees that would root in this boggy ground. They enjoyed the muck so well, in fact, they often grew in clumps, like bundled stalks of celery, each trunk a yard thick at the base. When the quarriers lived here, before the creek bottom was lumbered over, some of the hollow granddaddy sycamores were roomy enough for stabling a horse or—in a pinch—a family. Cross-sections of the solid ones, cantankerous and unsplittable, provided giant wheels for ox-carts. The upper limbs of the humbler trees we saw were laced together with grape vines. Meanwhile the flowers were putting on their May extravaganza: blue of waterleaf, pale rose of wild geranium, white of mint and false Solomon's seal, scarlet of columbine and fire pinks. Knee-tall reeds and horsetails, pulpy plants straight out of the Jurassic swamps, pushed up between the flowers. Had there been a brontosaurus in the neighborhood, it would have felt at home. Old-timers call horsetail the scouring rush, because its cell walls contain silica and a fistful of it makes a dandy pot-scrubber. The ancient ancestors of this spindly plant grew to a height of sixty or eighty feet. Sunk in the swamps three hundred million years ago and baked in the slow oven of the earth, they formed the coal that runs the generator that makes the electricity that powers the machine I'm writing on this minute. Coming on the lowly horsetail reminded me of the pertinacity of plants. Here they grew, frail jointed reeds that were older than the creek, older than the whistling birds, far older than two-legged photographers and writers, nearly as old as the limestone.

High waters closed the statehouse quarry after only two years. The blocks had to be hauled across the White River to the railroad, and floods kept taking out the bridge, so after a couple of wet springs the site was abandoned. Now all we could see of the settlement were the stone piers, watched over today by a pair of garter snakes, snazzy green with yellow racing stripes.

Because it lay across the White River on the railroad side and because it yielded sixty feet of good stone, the quarry at Romona—the people around here pronounce it "Roll-moany"—was kept open for a century, from the 1860s until the 1960s. "What finally shut it down," according to a farmer we met there picking up stones in a field, "was these two men who owned it died and the widows got so old they didn't know Monday from Sunday. Then one of the widows died and the other went crazy. They had to put her away. And she wasn't hardly out the door before the relatives got to fighting over who owned the place, and next thing you know the lawyers got into it, and now there it sits, a world of stone and nobody working it."

Somewhere in his fifties, filling his chocolate-colored T-shirt to more than capacity, the farmer was a glowering man who gave off an aroma of bitterness. The field he was standing in had just been cleared, and the bulldozer he had cleared it with, still ticking from the effort, hulked there beside him. Stumps and brush and weather-gray boards were heaped along the edges, ready for burning. A wagon hitched behind a John Deere tractor was parked within throwing distance. Every now and again as we talked, the farmer picked up a fist-sized rock from the dirt and heaved it clattering into the wagon. We had stopped to ask him about Romona quarry, whose derricks we could see rising from behind a screen of trees at the back of his field. Would he mind if we went to have a look at it?

"Damn right, I'd mind. What the hell you want to see it for?" He squinted at us with eyes like disks punched from an iron skillet. "You want to break your legs? You want to break your necks? You want to get yourselves *killed*? God *damn* it all to hell. Every time I turn around, honey, there's somebody sneaking back there fishing or necking or hunting or just fucking around. Son of a bitch go in there and break his neck, and the insurance will ruin you. You ever buy insurance? It's sky high, honey, let me tell you."

From all of this I concluded that he owned the land.

"Naw, naw," he said. "This bunch out of Ohio owns it now. All they use is the gravel. You want to go look at it, you got to get their say-so."

"In Ohio?"

"Naw. They got an office in Spencer." He chucked a lump of stone at the wagon. Bang, clatter. "I'm just clearing this field for them. Get it in corn, put it to use."

I looked at the weathered boards in the burn piles. "What was here before you cleared it?"

"Bunch of them shacks the quarry workers lived in. Little bitty things you couldn't swing a cat in. Rat-traps."

Jeff and I exchanged a grimace. We had missed seeing the old cabins by a few hours. "Did you work in the quarry?" I asked.

"Hell, yes, I worked there. Twelve years, starting in 1948. I was just married, you know, and pissing vinegar for a job. A cousin told me to come out and he could get me on. Shit, I told him, they ain't got any work out there that'll hold me. And shit, honey, inside a year I was right up at the top, and I stayed there till the quarry shut down."

Here I drew another false conclusion. "At the top? You were ledge foreman?"

"Shit, no. They didn't have sense enough to make me foreman. What I got to the top of was the derricks." It was his job every week to oil the gears on the hoist engines and the huge black wheels, wide as truck tires, on the stiff-leg derricks. The wheels on the boom he could oil from the ground, but the wheels on top of the mast he had to climb to, a hundred feet up a swaying steel tower. "You talk about fucking *dangerous*. You get on top and the wind's blowing. Nothing to hold on with but one hand, and that's greasier than pig fat. Nobody else but me would do it. I saw an oiler fall from up there once. He screamed all the way down the derrick and six floors down into the quarry. When we picked him up he wasn't anything but a sack of loose bones."

This led the farmer into a vehement recital of other quarry deaths. "Once this old foreman was standing between two blocks on the grout pile, honey, and the whole thing shifted, and his bottom half was squashed so thin you could just about see through it." A show of thickness, crusty palms half an inch apart. "We dug him out with the derrick and drove him up to the hospital in Greencastle, him moaning all the way. But he was deader than a hammer when we got there." Another time a drill-runner sat leaning against a tilted block to eat his lunch. "When it fell over, he never knew what hit him. Never drawed a breath or made a peep. Shit, honey, I'm telling you a quarry's a damn dangerous fucking place."

For half an hour he fumed on about men electrocuted when channelers cut through wires, men smashed by derrick hooks, men ground up in machinery, buried under mudslides, scalded in explosions, drowned in trenches, men whose hearts burst from lifting or hammering. "They never made a nickel off that hole after all them poor sons of bitches got killed. Even without them two widows taking it over and going crazy and all the lawyers fighting over it, the quarry was fucking done for." He swore randomly, cramming his sentences full of angry words. Yet even while swearing, and glaring at us with his skillet-hard eyes, he kept calling us "honey," the way my Mississippi kinfolk used to do.

I eased him onto the subject of corn, then soil, then fish, hoping he would relent and say we could go on ahead and have a look at the quarry. But when I asked him again, he snarled, "Hell, no. I won't have your blood on my head. First thing I know, honey, one of you'd break your damn leg and the other'd break your damn neck, and there'd be hell to pay with the insurance."

So we left him there in his bulldozed field picking up rocks, which he flung at the wagon with the furious determination of a relief pitcher nursing a one-run lead in the last of the ninth.

After half a dozen phone calls, a cautious official at American Aggregates finally said yes, we could explore their quarry at Romona, if we promised not to sue. The day we drove back out there opened with a thunder shower, and continued misty and cool. The farmer wasn't in his field. We parked down the road a piece, where the old railroad switch from the quarry joined the main line. The bulldozer had been at work here, too, ripping a muddy track through the woods for logging trucks to use, clearing an old house site. The house itself must have burned. A twisted ironing board and melted television cabinet lay inside the foundations on a bed of cinders.

The bulldozer had shoved to the back of the clearing a tiny shed, its joints awry. The roof was covered with black shingles, the walls with yellow asphalt siding meant to look like brick. Where the siding had peeled away, paintless boards showed through. They reminded me of the oak planks on a pigpen my father used to have. The shed, in fact, was about the right size for three prime hogs lying side by side. But the last occupants had been human, for inside I found a bottomless kettle, an oil heater, a burst shoe. I couldn't stand up in there, and neither could anyone else who's taller than about five feet. In the lone, glassless window a rag of lace tablecloth hung down for a curtain, sashaying in the breeze. Behind the shed was a one-holer privy. Around the fertile foundations grew clusters of white star-of-Bethlehem, nodding timothy grass gone to seed, and waist-high ferns.

"Do you think some of the *quarriers* lived here?" said Jeff.

We had no way of knowing for sure. Neither of us liked to think of *any*body sleeping there, least of all a man who staggered home bone-weary after ten hours of muscling stone.

We staggered to the quarry through the woods. I led the way. Trusting my sense of direction, Jeff labored along behind with the tripod and camera over his shoulder. By way of rutted logging tracks, ravines, bramble patches, and wet thickets, I did eventually guide us to the Romona pit, only to discover there two fishermen, fat as Buddhas, who told me we could have sauntered in along a broad cinder path. Jeff commented on my orienteering skills. I asked the fishermen about their luck. They'd caught only a single bluegill, but—Lord, Lord!—they'd seen bass as long as your leg, bass so big if you ever caught one you'd need a cart to haul it home, bass of mythical strength and cunning. We listened skeptically. Then we trekked down to one end of the huge pit and peered over the edge. And Lord, Lord! There were bass straight out of fairy tales, as long as a forearm if not a leg, their bodies dark shadows against the pale rubble, idling in the murky water like apprentice sharks.

The quarry itself was on a scale worthy of the fish. After a century of digging, men had left a trench broad and long enough to hold ocean liners. At its deepest, from ridge line to the underwater bottom, the drop must have been a hundred and twenty feet. Only a serious swimmer could have made it from end to end, and would have had to navigate around peninsulas of cattails, where red-winged blackbirds were fighting for territory, and past archipelagos of tumbled blocks. In this great hole the reverberating voice of bullfrogs sounded like diesel trucks shifting gears. The stairstepped walls, rising story after story, looked like the ruins of a Mayan temple.

Three derricks had been left in place at one end of the trench. Their supporting cables intersected overhead in a spider's web. A kingbird and a crested flycatcher hunted insects from the guy wires, flitting out for the kill and then perching again to survey the gulf of air. The steel masts bled rust onto the rock below. In paint even brighter red than the rust from the derricks, somebody had scrawled the word LIFER here and there on the quarry walls. Everywhere the word appeared, it had been peppered with rifle fire. In some of the pockmarks the bullets were still visible, buried up to their midriffs in stone.

From the ledges, hundreds of yuccas thrust up the green swords of their leaves, and the red blossoms of columbine drooped like the falling sparks of Roman candles. On our way back down the cinder path to the car, we surprised a gray fox, which took off loping ahead of us at a dignified pace, back and tail undulating, like a dolphin cutting water. It was not the first or the last of the foxes we saw around the old quarries, but it was unusual because of its color—red foxes prefer stony ground, gray ones prefer brush and woods—and because it showed itself at midday. Usually we saw foxes at the edges of the day, early morning or dusk. They would lift their muzzles and prick up their ears when they caught a whiff of us, then give us a long, curious stare before trotting off with the smug air of aristocrats. Coyotes also make their lairs in the old grout piles, we were told; but they're shrewder even than foxes, and we never met with any. Back at the car, eating our cheese sandwiches and apples, we were visited by a ruby-throated hummingbird. It rested for a minute on a telephone wire, then zipped away, headed for a drink from the red fountains of the columbine.

At midsummer, on the stone lips of abandoned quarries, columbine is replaced by the tough open-field wild flowers: the luminous orange of butterfly weed, the creamy bursts of multiflora rose and the white tracery of Queen Anne's lace, pink everlasting pea, purple joe-pye weed, red clover and bull thistle, violet garden phlox, lavender vetch, blue chicory. They make quite a show in the dog days, a garden no one plants and no one weeds. The high summer sun also deepens the green of quarry water. On the topographic maps these wet holes appear like scars, as if the land had been ravaged by smallpox. Zooming over at five hundred feet, jets from Grissom Air Force Base use the belt of quarries for navigation, flying up the seam of glinting scars. On foot you swelter across green fields and stare over the edge into green pools, field blurring into quarry. Algae blooms down there like brown clouds, confusing the issue. Sky and underearth meet on the shimmering surface.

Fish eggs ride into the water-filled quarries on the feet of ducks. Bluegill, crappie, bass. Swimmers ride there on bicycles and motorcycles, in pickups and convertibles. The last few hundred yards they'll walk, if the water is clear and deep enough. Afraid of stony silence, they come with radios. To domesticate the raw pits, they bring air mattresses, coolers of beer, footballs, crinkling sacks of groceries; they park their vehicles within easy reach, ready for a getaway, like cowboys afraid to lose sight of their horses. They do all they can to transform the quarries into backyard swimming pools.

Near Bloomington a rusty-bearded man named Rob Titlow has spent eight years loading mail trucks at night, studying law in the daytime, and on weekends turning one of the quarries on the land he rents into a perfect swimming hole. He picked out this cratered land from a helicopter, after surveying all the quarry country for a likely place. "I didn't grow up around limestone," he told us. "But when I finished with Vietnam and came here to school, I got to where I loved the overgrown grout piles and the wet holes and the derricks sticking up against the sky."

He'd like to turn his two hundred acres into a wildlife refuge, if he could get together enough money to buy the land. "We have beaver, foxes, coyotes, quail, pheasant, and so many deer it makes my buddies at the post office cry just to think about it. But I don't allow any hunting here." Fifty of those two hundred acres have been quarried. "When the stone companies finished with a pit they'd dump it full of trash—cables and channeler rails and broken machinery—to discourage swimmers. I chose the prettiest hole and went diving in it and cleaned out every bit of junk. Truckloads and truckloads. Then I stocked it with bullhead minnows to get rid of the algae. They're like vacuum cleaners, and they won't nibble on you the way bluegill do. Now it's about as fine a place for swimming as you'll find. I try to keep everybody out except my friends."

Titlow's circle of friends must be large: on the day we visited, there were nine cars and two bicycles parked at the head of the path leading to his perfect swimming hole. All guests observed two unwritten rules: to leave one another in peace, and to wear no clothes. "The guys who work at the quarry that backs onto my property like to park up on the ridge and watch the women. I hate it, because the reason my women friends come out here is to get away from the whistlers and gawkers. But I can't really hassle those guys. If they got mad and dumped a barrel of oil in my quarry, the place would be ruined for swimming."

The ordinary wet holes, which no one has groomed for swimming, are lethal places. Almost every summer a few swimmers are hurt or killed. Diving from the ledges, a man cracks his skull on rubble hidden beneath the soupy surface. A girl paddles into the mouth of an underwater cave and never paddles out. Grabbing a wire, perhaps thinking to swing out over the water and play Jane of the jungle, a woman is electrocuted. A boy, while frog-kicking along the bottom, gets caught in the submerged skeleton of a derrick. Owners post warnings. Newspapers run horror stories. But the swimmers return every summer with the stubbornness of migrating caribou. A cashier in a plumbing store told me that when she was in high school she and her friends once found a dead cow floating in their favorite hole, but they just pinched their noses and jumped in anyhow. Summer after summer the sheriff's men make raids, sometimes arresting the swimmers by dozens; but they do so with half a heart, because most of the sherriff's men, when boys, swam in these same treacherous pits.

The old quarries are doorways, opening down into elemental depths. They seem to inspire elementary passions in people, or to attract people whose primal urges have already been roused. Within the space of a month, rambling on my own, I witnessed three scenes that have stayed with me.

A red pickup, so heavily loaded that its rear bumper was nearly scraping the ground, backed to the edge of a roadside quarry. Two women hopped out and peeled a tarp away from the bulging cargo, which appeared from where I sat to be a small household's worth of furniture. Judging by their angry conversation while untying the ropes and by their matching cheekbones, the women were mother and daughter. The older one barked orders, the younger one sullenly replied. The mother was about forty, husky, as quick on her feet as a boxer, hair rolled up in a brown beehive. The daughter was in her middle teens, skinny, languid, face shadowed and hair hidden beneath a railroad engineer's cap. When the last rope was undone, the mother seized a rocking chair and heaved it over the lip of the quarry.

"I'll teach that goddamn son of a bitch!" she yelled. She kept on yelling about a man whom she never named as she threw stools, lamps, tables, cardboard boxes stuffed with clothes, armload after armload over the side. The stuff landed with a sploosh or a muffled thud, depending on whether it hit water or stone. Every now and again the woman hollered at the girl to help, but the daughter hung back, hands stuffed in jeans, face tight.

"He can't treat me that way!" the mother shouted. Over went a small television, the drawers of a rickety dresser, a pair of chrome-armed kitchen chairs. "Get your hands out of your pockets and *help* me," she cried as she wrestled with a sofa. The daughter never stirred. The sofa tumbled over anyway, and after it a mattress, bedstead, rolled carpet, boots, and a Sears catalogue.

Eventually the truck was empty. The mother had never ceased yelling at some absent and beastly man. The daughter had never budged from her sullen pose. I watched it all from thirty feet away, squatting in the sun beside my parked bicycle, and all the while neither woman took any notice of me. Then as she slammed the tailgate, the mother glared at me and hollered, "And you can go to hell!"

Doorways into the Depths

I sat tight, mum, not wanting to join that other man's household goods down in the bottom of the quarry. The pickup peeled gravel when it left.

Another evening, and in another quarry, I was sitting on the steps of a caboose. When this pit was yielding up stone, the caboose had served as foreman's office. Now it served me handsomely as a perch, from which I could study the darkness thickening among the limestone ledges. A shadowy figure broke the skyline and walked past me to the lip of the quarry. It was a man, rather stooped, and he carried something rolled up in his hand. I could make out nothing more definite in the darkness. There was the click and flare of a lighter. In a moment, whatever the man was grasping became a torch. He lifted it up near his eyes, giving me a glimpse of an old man's face, wrinkled as a dried apple, solemn and impassive. Was it a will? I wondered. A marriage license? Bundle of bills? Love letters? Poems? And why had he brought it here, instead of burning it in the fireplace at home, or in his driveway? When the flames were guttering down near his fist, he dropped the torch onto the stone. In a slow shuffle he circled the fire. Once, twice, three times. On each drowsy orbit his cratered face waxed and waned, moonlike, in the reflected light. It was a primordial dance, a banishing of some evil. Whatever the source of his dread, it required this stony place for its exorcism. After the last spark winked out, I heard his shuffling steps go by the caboose, almost near enough for me to reach out from my cloak of invisibility and touch him.

In the third and last of these little passion plays, I was planted like a seed in a waterworn cranny high in the wall of a quarry. Of course it was a womb that held me. But it was also an ancient seabed, a good place for sinking down into the depths of the mind. The day was hot, crazy hot. A red-tailed hawk circled in lazy loops overhead, waiting for anything with blood in it to show itself on the rock. An old wooden water tower, held together by rusting steel bands, leaked a constant stream onto the ground nearby. I was reflecting aimlessly on hawk and water and stone when a van pulled up. On the side a desert sunset had been painted in day-glo colors. Three men clambered out, all in baseball caps advertising beer, each with a can of beer clamped in his fist, each a bit loose in the joints of the legs. One of them caught my eye in particular, for he was remarkably fat above the waist and scrawny below, like a keg balanced on a sawhorse, and to show off his physique he wore the sort of belly-length T-shirt favored by svelte tailbacks.

This one led the way to the quarry lip, just around the corner from the niche where I was planted. He jammed the beerless hand into his pocket and drew out a revolver. Bam bam bam. He fired down into the water—at no particular target, so far as I could tell from the scatter of splashes. His two buddies also tugged handguns from their pockets and began pumping lead into the green pool. The noise hammered around in the quarry like a maniac in a padded cell. Maybe they were aiming at the fish, which veered in silver bursts, scaly sides catching the light like a slant of wind-driven sleet.

The trio emptied their guns, reloaded, emptied them again, three times, and on the fourth round they began firing at the walls. They laughed, hearing the bullets zing and ricochet off the stone. I was not laughing. I had crawled so far back into my cranny that it would take a good lubrication of sweat to get me unstuck. They might have killed me by accident. But they might also, I was convinced, have killed me on purpose. It would have seemed more in keeping with their helter-skelter mayhem to have shot me than not to, and nobody could ever have known they did it. I kept well hidden. A quarry in a quarry. Except for their laughter, not a sound emerged from their throats. They spoke only in bullets.

By and by the shooting stopped, the van's motor started, the tires crunched away. I was a long time in coming forth.

5 A Veteran

Of all the husky machines for digging stone, my favorites are the derricks. Bulldozers are commonplace. You can see them nosing around any day of the week, in any township. Channelers are unique to the quarries, but they seem plodding, dimwitted, like anything that shuffles back and forth in a rut. The pneumatic drills are cocky but puny, almost comical in their rattling assaults upon bedrock. They remind me of Joseph Conrad's account of French gunships firing salvos into Africa. Pow, pow—surrender.

No, give me derricks. They are the first things you see as you approach a cluster of quarries, the upright beams thrusting skyward like the masts of a fleet at anchor. Guy wires, slanting down from the tops of these masts to bolts embedded in rock, hold them in place. These webs of cable summon up memories of carnivals and circuses, as if roustabouts might hang a tent on each of the brawny poles. It takes a dozen men more than a week to take one down and set it up again in a new spot. By means of a boom, which is hinged at the base

A Veteran

of the mast and controlled by winches, the derrick can sweep out a circle two hundred feet in diameter and heft forty-ton stones as if they were children's building blocks. After the chisels and drills have finished their brute work, the derrick performs the delicate business of lifting and stacking. A glimpse of the limestone leviathans dancing against the sky on threads of steel is reason enough for admiring derricks.

The puppeteer who makes the stone dance is called a powerman. He sits on a high chair in a shack of corrugated steel, pulling levers that resemble the handles for emergency brakes on old trucks. In front of him there is a nest of gears, like a microscope's view of the works in a watch, and on the floor at his side there is a five-gallon can of grease, with a long-handled dipper for ladling goop over the cogs and wheels. Drive cables as thick as a woman's wrist stretch between the power shack and the derrick, making the contraption go. Most of the time the powerman can't actually see the stone he's lifting. Instead, he's guided by hand signals from the derrick runner, who stands on the rim of the quarry and is guided in turn by signals from hookers down in the pit. A touch on the helmet means raise the boom; palm out like a policeman—stop; wrist bent down sharply—lower cable; twirling of wrist—rotate mast; palms pressed together as if in prayer—raise boom and lower cable. And so on through a brief lexicon. It is a simple speech, and an exact one. A misunderstanding could ruin a block worth thousands of dollars; a slip of the hand or eye could kill a man.

"If you're looking for an old powerman," Jeff and I were told at Summitt's Grocery in Stinesville, "you drive out Mt. Carmel Road and see Cedric Walden."

STONE COUNTRY

Unsure how far out Mt. Carmel Road to drive, we stopped to ask directions from a boy who stood as if posing for a postcard beside a mailbox, a wand of timothy grass projecting from the gap between his front teeth. "Mr. Walden? Yessir," the boy said, pointing, "he lives right up over that next hill in a little old bitty house, and he's got a horse and a tractor and a motorcycle and a truck and I don't know how many dogs."

The horse, a frisky palomino, snorted and stamped in a muddy lot beside the house. The motorcycle was out of sight, but a vintage tractor and pickup sat in the gravel drive, their color a compromise between red paint and rust. A well-clawed dog cage filled the back of the truck. The house was little bitty, all right, a one-room box of peeling clapboard, with horseshoes nailed over the front door.

Walden greeted us there under the horseshoes, a mug of coffee in his hand and a grin on his face. "How are you doing?" we said. "About all right," he answered. "We're interested in limestone," we said. "In that case you better come on along inside. I don't know a whole lot about many things, but I guess I know as much as anybody about getting stone out of the ground."

The room held an iron bed with a checkerboard quilt, a television, a vacuum cleaner, two heaped bureaus, two straight-backed chairs, and a brand-new wood stove. Because the morning was chilly, Walden had ignored the fact that we were in the last week of May and allowed himself the luxury of a fire. An open toolbox, gleaming with wrenches, rested on the floor beside the bed, and beneath the bed were two scuffed helmets, one for the motorcycle, the other a miner's hat with lamp. "That's for hunting coons at night. You just switch on that light and then you got both hands free for your gun." The guns, three of them, leaned in the corner behind the door. Aside from the Bible, the dozen volumes scattered about on the bent-wire TV stand

A Veteran

and dressers were all *Reader's Digest* condensed books. In the hazy windows on the south wall, there were amber bottles and two coondog trophies. We heard the dogs out back, their voices throaty and melancholy like a nightclub singer's. "Two blue ticks and a black and tan. One of them's still a pup, and another one's no account. But the third one's got a real good nose." The room smelled of pipe tobacco—a can of Prince Albert rested on the floor beside Walden's feet—and of bachelorhood.

Toothless, lips drawn in as if his mouth were the socket for a third eye, a week's white stubble along his jaw, his face more lined than a road map, his body stiff and worn down to a frazzle, Walden had lived his seventy-one years pretty hard. But he still burned with the high spirits and gumption of a young man. He had bought the motorcycle six years earlier to use for visiting a ladyfriend in Cleveland. "That's ten hours away, and you got to go through cities to get there. Indianapolis, Columbus. I'd never been on my cycle anyplace bigger than Spencer, so I rode down to Bloomington for a practice, and I was near scared out of my wits by all the traffic. People just drove any old ways. Whizz! But I put my head down and kept riding, and soon I got over it, and I made it to Cleveland all right. I liked it so well I did it five times, and went to Florida, too, and on out to California. Only trouble I ever had was I come all the way back from Denver once and had a flat tire three miles from home."

He sat on the edge of the bed, his back as straight as the iron headboard, slicing the air with his left hand as he talked. The right hand, clasping an unlit pipe, lacked an index finger. "I got hurt plenty of times in the quarries," he explained, "but that wasn't how I lost that finger. How I lost it was, a Model-T Ford fell off a jack and cut it clean off." He wore ankle-high boots, unlaced over sockless feet, Oshkosh bib overalls, a faded cotton shirt, and a baseball cap with a picture of a Piper Cub and the inscription: WALDEN'S FLYING SERVICE, HASTINGS, FLORIDA. "My brothers are cropdusters down there. They made it real good." His father and grandfathers had worked in the quarries, but his brothers and sisters left the stone belt, and so did four of his five children. One son kept the family tradition alive by working in a mill near Bedford. "That boy picks things up like a top. You only have to tell him a thing once, and he knows it."

At fifteen, on the brink of the Depression, Walden earned his first paycheck at a furniture factory in Bloomington. "Sixteen cents an hour. Do you believe it? I thought I'd hit the big money." Even then he wanted to work with stone. Between about 1932 and 1946, however, there wasn't enough business in the quarries and mills to keep even the veterans busy, and there were no places for new men. When building picked up again after World War II, he went to quarrying, and he kept at it for over thirty years, until his retirement in 1978. "I did everything there was to do. Run the channeler, drilled, hooked, stacked. What I did best and longest was run power. It's all done by signs, you know." With lumpy hands criss-crossed by scars he ran through the gestures swiftly, like a third-base coach giving signs to a batter. "When you're running power, you can't have a thing else in your mind. You can't go thinking about fishing or women or food. You got to keep your eyes and ears open. The power's got a sound to it. All the wheels and cables, the engine, everything together makes a sound. You get to know how it ought to be, and you can tell pretty quick if it ain't right. Every morning before I started up I'd check over the works, you know. Once I missed seeing a crack in a wheel, and along about noontime it busted, come out of there and drove me right back out of the powerhouse. Broke my collarbone and laid open my head. If I hadn't been leaning forward on the lever

A Veteran

when it busted, I wouldn't be alive to sit here and tell you about it now."

Another time, when he was working down in the pit as third hooker, "One of the hooks tore loose from the doghole and come at me like a bullet. All I had time for was to get my hands up in front of my face. Couldn't duck, couldn't run. Just stood there and had that dog beat me on the head, three times. I laid out a week or two after that. When I went back I told the foreman I didn't want to go hooking for a while. I didn't have the nerve to keep near the stone just yet. And if you ain't got your nerve around stone, you ain't much good for nothing."

Now and again government officials would come to advise the quarryworkers about safety. "Half of them never been in a quarry before. They looked at my powerhouse and said, 'You got to cover up all them gears with boards.' Boards! Cover them up! When all that keeps you alive is watching that power work and listening to the sound it makes. Why, cover it up, and you're as good as dead. Some of the owners I've worked for don't know much better than the government inspectors. They're smart, all right, and know how to keep the books. But when they come to sticking their nose in a quarry most of them are lost sheep."

Over the years he had worked in nearly all the quarries in the district. "If things didn't suit me one place, I'd move on to the next. You had to work five years straight for one outfit before you'd get a two-week vacation. There for a long while I never could stay put anywhere long enough to earn me a vacation. I'd get mad and quit and take my lunch bucket someplace else. Anybody'd hire me, because they knew I was good." What made him mad? "Well, you take this foreman, Carl Penny. He was the worst man to work for I ever see. He didn't want you to sit still a minute. If you finished everything there was to do, he'd *make* work, just so he wouldn't have to look at you sitting still. One time he had us to move a stack from one side of the hole to the other, and then move it back again. Well, I don't put up with that kind of thing. Penny'd make me so mad I couldn't see straight. He'd get on me, and I'd pop off at him. Then he'd get on me some more, and then I'd balk on him. Next thing you know, I was carrying my lunchbox out the gate. There's been more than one time I've quit one place at eight and been hired on another place by nine o'clock, and my lunch already packed."

Moving about, he once drove down to see how he would like digging limestone in Texas. He didn't like it much. "That stone they've got is so soft, it's about like chalk, and the place where you've got to live if you work in the quarry is right smack dab in the middle of nowhere." After a week of Texas he headed back to Indiana, leaving behind a daughter—who is still there, working as a nurse. "A nurse is like a stone man used to be. Anyplace you go, they'll hire you on the spot."

On the side he farmed a little, not much, some corn, some alfalfa. "You work a day in the quarry, and you ain't got much juice left to do anything else." In winter, after quarrying finished for the year, sometimes he stayed on to help load blocks out of the stack. But usually he went home and collected unemployment checks. "Most of the boys in the quarry felt the same as me. I like working outside, and I like having my winters to loaf."

His favorite manner of loafing was to chase raccoons through the woods all night behind a brace of good dogs. "When my legs began to give out and I got to where I couldn't hardly *go* anymore, that's when I bought my horse. Her name's Lady, and ain't she a beauty? I figured I could ride through the woods if I couldn't walk. But them boys I'd go with went chasing coons through places in the pitch dark where I wouldn't go at noonday. *Wild* places. I got to thinking, 'What hap-

A Veteran

pens if I fall off away back in there and can't get back on? *Then* what kind of a fix will I be in?' So horse riding didn't pan out, and I quit hunting. Burned me up. I got one good dog left, but another man hunts him." Mounting with the help of a fence rail, he still rode the palomino between house and town. "When she breaks out of her pasture, I go after her in the truck. Once my neighbor down the way led her back up here with four children aboard." One of the four, we discovered, was the gap-toothed boy who had given us directions to Walden's place.

In winter now, unable to hunt raccoons, he watched Indiana University basketball games on television. "I get so worked up, I can't hardly stand it. If you was to see me you'd think I was crazy, the way I carry on, out here all by myself hooting and hollering over a ball game." And in winter he fed the birds, whose spring whistling accompanied our talk. Robins had nested this year in the toolbox beneath his tractor seat, so he never started the motor. When a gust of wind tipped over the nest in March, he picked up the eggs and replaced them. He was afraid the mother would reject them, but she didn't. And now she would take bread almost straight from his hand. "But she don't give it to the chicks until she's carried it off somewhere in the woods and sniffed it over, to make sure it's all right."

Out front again, he pointed at the huge ash tree next to his drive. "That old boy don't have long to live. It's been struck by lightning three times." Would he cut it down for firewood? "Not me, sir. Nowadays I buy my wood. I can't run a saw, can't hunt, can't hardly get on a horse. Ain't worth a nickel." He cocked a rheumy eye at us to see if we would contradict him. Despite the toothless mouth and the bristles of whiskers and the bib overalls, there was about him the dignified air of a judge. He knew his worth, and didn't need any visitors appraising it for him.

6 Poison

Wanting to explore Hunter Valley, a rash of quarries on the north edge of Bloomington, Jeff and I drove out to ask for permission and a gate key from the owner, Ed Bennett. A bypass had carved up the valley, baring prime stone in the roadcuts and scrambling the map, and we were soon lost in a snarl of gravel drives and snaky access roads. We stopped to ask directions at a rickety tarpaper hut that appeared to stay upright by leaning into the wind. Holding five cats on his lap, an old man with a face like a hatchet sat rocking on the front porch between a refrigerator and a cement mixer.

"If I was you I wouldn't have nothing to do with Ed Bennett," he told us. "Go hunt you up a pole cat to talk with before you go talking with Ed Bennett."

"You know him?" Jeff asked.

"He's my cousin. I've been knowing and fighting him for eighty years. What do you want him for?"

Poison

Jeff said the word, the open sesame for loosening lips in this country.

"Limestone?" this amiable cousin repeated. "I'll tell you about limestone." Then, without further provocation, he unraveled a yarn from his youth. Sometime in the twenties, when he also was in his twenties, he was driving up to Spencer one Saturday night for a dance at the Union Hall. It was about sundown, and he was all slicked up and primed for pleasure after a week's drudging in the quarries. And what do you know but halfway there he saw a beautiful girl dressed in a white gown hitchhiking beside the road. Naturally, he picked her up, took her to the dance, and along about midnight proposed to her. She said yes. But first she had to go back home and break the news to her folks. So after the dance he drove her way out in the boondocks somewhere, way to hell and gone back in the country there round about Mt. Tabor, and let her out at her place, and promised to come get her the next day. On Sunday morning he drove back to her house. In the daylight it looked all tumbledown. He knocked, and this old fagged-out woman came to the door. He took off his cap and spoke the name of the girl he was supposed to marry. The old woman scratched her chin, thought a minute, and said, "The only place I ever come across that name is on a tombstone out back." Then she took him into the yard and showed him the grave.

"And what do you know, but that girl I waltzed and smooched with had been dead for near about a hundred years!" Tying up the knot of his story, Bennett's cousin sat rocking in the niche of shadow between the refrigerator and the cement mixer, petting his lapful of cats.

I had been told the story of the lovely hitchhiking ghost before, but never with so much melancholy as this hatchet-faced old man had squeezed into it. The link with limestone was too faint to make out. We reminded him that we were looking for Ed Bennett.

"What do you want with that old pole cat?" he demanded, as if we had never mentioned the name before. Then he waved us directions with ropy arms.

A quarrier for nearly seventy years, Ed Bennett lived next door to a limestone house, but his own place was frame. He met us at the door in blue work clothes, green suspenders, and beat-up leather boots spotted with red paint. He'd been out cutting grass and his left hand was bandaged from a run-in with a pulley on the tractor. The top of one ear was scabbed from a recent brush with the engine on his car. When he held up his good mitt for a handshake, the scars and lumps showed that he was a man who had never babied his body. He had a good deal of body. His chest and belly formed an unbroken curve that resembled, in profile, the bulge of a nail keg.

Like most of the old stone men, Bennett was a little suspicious about why two pointy-headed young guys from the university should be so interested in limestone. He lowered himself into an overstuffed chair in the living room and eyed us up and down. There was a magnifying glass on the table beside him and a brass spittoon on the floor. Every now and again, as we spoke, he leaned over and let out a spurt of juice. He had additional reasons for being suspicious just now. In the 1970s he had permitted the dumping of some electrical capacitors in one of his quarries, and the fluids leaking from those capacitors were poisoning the neighborhood with polychlorinated biphenyls— PCBs—recently added to the Environmental Protection Agency's hit list of deadly chemicals. In the past few months the lawyers and scientists and politicians had pestered him near to death.

"I never even *heard* of any PCBs until these legal letters started coming in the mail." He riffled a stack of envelopes on the side table. "The print's so small you need a glass to read it." Since he was leaning over the arm of the chair, he let go at the spittoon. "How was I to know them capacitors was poisonous? And now they're saying I tried to hide them. Well, hell, I pushed them around with my bulldozer, but I wasn't trying to cover anything up. I was just leveling out the ground. If I'd wanted to *hide* them, why, in half an hour I'd have buried them so deep nobody'd *ever* find them. But these officials have got to yell about something, and right now they're yelling about some chemicals in my quarry. If you ask me, it's all a bunch of hooey."

As we asked him questions about Hunter Valley, Bennett's manner became less defensive, and soon he was peppering us with stories. "The Hunter family they named the place after had a boy name of Johnny. And he was a scamp. There came a time when Johnny's momma got to missing wheat out of her granary. Every time she'd look, there'd be less of it. Where'd it go? They kept a lookout for rats. But what it turned out to be was, Johnny'd taken him a brace and bit, bored some holes in the floor, and he was draining out the wheat and selling it. Stealing his own momma's wheat and selling it for pocket money. That's the Hunters for you, right down to the ground."

Bennett's father ran a stripping outfit, clearing overburden for new quarries. Bennett himself started in that business, then bought thirty-six acres of land in Hunter Valley, then piece by piece bought up the remaining three-hundred-odd acres of the quarry field. He knew every inch of it.

"There was five or six company houses down next to the Vernia Mill. Some dagos was living there, you know, stone cutters they brought over from Italy. And in Prohibition time they made beer and sold it around. 'Blind tigers,' we called the stuff. You could drink it all night and it wouldn't make you sick. It was that good. The boys

STONE COUNTRY

would drive their teams and wagons by and load up with that Italian beer. Twenty-five cents a quart. Anywhere you go in those quarries on a Sunday, and there'd be men sitting around with sixteen-gallon kegs and tin cups, and they'd drink till the beer ran out. Then along come Monday, and it was back to work."

A hiss of running water and a clatter of dishes sounded from the kitchen. Bennett hitched up his belt, gathered a mouthful, spat.

"There was hermits living back in them quarries all during the Depression. Still are today. Hobos, you might say. Tramps. They'll get themselves an old sheet of tin for a roof, wrap up in newspapers to keep warm. Some of them are barefoot. Never talk to a living soul. The boys put food out for them, you know, just like you'd put out food for a dog, and in the winter they'll leave the stoves burning in the mills to keep the hermits from freezing."

The clatter of dishes rose to a crescendo and then stopped. A white-haired woman—wispy and slim, with the muted prettiness of a figurine from half a century ago—walked in from the kitchen, drying her hands on a calendar-towel. To Bennett she said, "Do you remember what year the Consolidated Mill burned down?"

"1921," he answered.

"Do you remember what month?"

After a moment's ponder, he said, "July or August."

She smiled, and for a moment there was nothing antique about her prettiness. "It was the last week of July, right about noon. I know exactly when it was, because we'd only been married three weeks. I was cooking with Momma, and we went outside to look at the fire. All I could think about was you being down there in it. My uncle was working on the traveler, and my father on the planer. But all I could think about was you."

Having made her personal connection to this history we'd been gabbing about, Mrs. Bennett slipped back into the kitchen.

Poison

Talk of the Consolidated Mill reminded Bennett of a photograph we ought to see. Curled and yellow, rolled up like a scroll, it was a two-foot-long panorama shot of the Consolidated crew, taken sometime back in the teens. Wearing aprons, the mill workers stared out at us from beneath the visors of flat caps. He laid a mashed finger on one of the proud faces. "Just before the mill burned down that poor son of a bitch got caught in a flywheel on the belt drive. In about half a second he was jerked straight up in the air and hung there by his suspenders." When Bennett paused to give vent to his tobacco juice—leaving the mill hand still dangling by those suspenders—I began to think the story would have a comic ending. But then he continued: "And I looked up and saw the top of his head had been sliced clean off and his brains was spattered all over the roof. A week later the mill burned down."

"An accident?"

"No accident. But I'm not going to tell you who set the fire, because the guy's son still lives down the road. If I'd name names, he'd be up here after me. That's Hunter Valley for you."

Among Hunter Valley's dozens of quarries—Half-Mile Hole, L-Hole, Star Hole—the fetchingest by far is one named Crescent. Roughly U-shaped, as its name implies, steep sided and ten feet deep in water, it might have been landscaped by a Zen gardener. On the first of May the wild cherry trees, gnarled and wind twisted like bonsai, were in delicate white bloom along the rim, on the tongue of backfill that stretches into the water, and on the blocks that form chunky islands across the miniature sea. Cedars flickering like green torches clung to the narrow ledges between floors. Higher up the bank, dogwood in blossom spread their creamy branches and redbud simmered with new growth like jets of purple flame. Last year's brown seedpods gave sumac the look of burned-out candelabra.

STONE COUNTRY

Jeff set up his tripod on a tippy slab that cantilevered out from the brink. "If this baby goes over the edge," he said, "save the camera." The weather and the place excited him. Everywhere he looked the shapes of things leapt and shouted. It was a gusty day, the clouds whipping past. "Great for photography," Jeff said. "If what you see doesn't suit you, wait thirty seconds and you'll have a change of light."

The quarry brimmed over with birdsong and the rustle of falling water. Cliff swallows cruised above the pool, giving off a steady grating, like the sound of pebbles dropping onto tin. After a few swooping runs they returned to their mud nests on the wall, gliding up before dipping to the roosts where the chicks waited with open, jabbering mouths. Turkey vultures circled overhead, flattened black V's as in children's drawings. Red-tailed hawks rode the thermals. Here and there springs gushed through cracks in the wall, spreading beneath them a tongue of moss. Near the bulge of the crescent, a creek tumbling over the lip had worn a path down through six feet of clay to the bedrock. Snails oozed about on errands beside the creek, leaving slick trails over stone inlaid with fossils of their three-hundred-million-year-old relatives. The black whorls of ancient mollusk shells showed up against the silvery limestone like dark spiral galaxies. In mud beside the creek, the cylindrical fossils of crinoids lay jumbled together with the cylindrical shell casings of .22 bullets. The skull of a possum rested on the bank with its snout in the water, as if it had taken one last drink and died.

Had the possum sampled the water seeping from Bennett's Quarry, up at the north end of Hunter Valley, it might very well have died from drinking. Among all the PCB dumps in America, the quarry named after Ed Bennett was one of the dirtiest and most dangerous. A year before our visit, technicians in breathing gear and bubble helmets and snowy protective suits, like astronauts on a hostile planet, had poked around, hauled away a few truckloads of leaking capacitors, filled the hole with fresh dirt, capped it with plastic, fenced it off, planted it with grass, and declared the three-acre site "stable."

The chain-link fence still gleamed in June when we drove out for a look. Topped by three strands of barbed wire, it was the same ominous barrier you will find encircling military bases and nuclear power plants. A man who'd worked on the fencing crew said that, when driving posts, they kept running into buried capacitors, kept moving the fence line farther out, kept hitting capacitors. Eventually they got fed up and returned to the original boundary. Inside, waist-high grass bristled on the fill dirt. In all likelihood grass and trees and the skittering insects would inherit this land forever, and the fences would never come down. I had the sense, looking in through the wire grid at this desolate patch of earth, that I was seeing the future. A derelict railroad spur ran along one boundary of the poisoned reservation. Yarrow and multiflora roses flowered between the rotting cross-ties, and wild strawberries were coming ripe. I was tempted, but did not taste these chancy fruits.

7 The Men in the Trenches

Limestone comes from the earth. Swallow some, and your body will know what to do with it. Unlike PCBs, DDT, TNT, and the other alphabet chemicals we concoct, which hang around to poison us after they have helped us, limestone is wholly benign. Grind it up, and it will fertilize your garden. Grind it even finer and you can stir it into your toothpaste or add it to your chicken feed. Unlike the radioactive tailings from uranium mines or the creek-killing debris from coal mines, the waste rock strewn around an abandoned quarry harbors no nasty secrets. Wait a spell and plants will throw a lushness of green over the ruins. Lichens, mosses, ferns, grasses, flowers, trees: underneath a blanket of roots the bedrock goes back to bed. Beasts will set up house in the grout piles, fish will claim the flooded holes, birds will nest in the rotting sheds.

But if you come upon a quarry soon after it's been abandoned, you'll think you've discovered a battlefield. Rock litters the floor and brim like rubble in a bombed city. The ragged pits might have been the basements of vanished hotels. Stones weighing tens of tons lean against one another at precarious angles, as if they had been hurled

The Men in the Trenches

there by an explosion. Wrecked machinery hulks in the weeds, grimly rusting—cogs and wheels, twisted rails, battered engine housings, trackless bulldozers, and burst boilers, like junk from an armored regiment. Everywhere the ledges are scarred from drills, as if from an artillery barrage or machine-gun strafing. Stumbling onto one of these raw pits you might be left wondering who had won the battle, men or stone.

In a working quarry, where the battle is raging right before your eyes, the outcome is no more certain. The pit is choked with dust and smoke. There is a fearful racket of engines growling, saws and drills whining, men shouting, chisels hammering. Forklifts the size of earthmovers lurch and grunt, butting their tines under blocks, like Babe the blue ox going after Paul Bunyan's trees. Overhead, suspended from cables, slabs of stone loom like boxy meteors. This is no place for the faint of heart. Like all mining, quarrying is brutal. Ripping into the crust of the earth, tearing hunks of it loose, and shipping the spoils away cannot be a gentle business. And who wins the battle? The machinery is powerful and the men sturdy, but the limestone resists stubbornly, with its passive weight and also with its hidden flaws, its fissures and color changes, its mud-filled seams and pockets of coarse fossils, flaws that render much of the stone unsaleable and so render the labor fruitless, a kind of defeat. (You are doing well if you can sell half of what you dig.) Quarrymen are not often braggarts, as some operators of heavy machinery tend to be. Their works are mighty by human reckoning, but measly on a planetary scale. They know that beneath their deepest digging, the rock goes down and down, and that even this stubborn shell is but a thin skin around the earth's fiery core.

The man easing himself earthward from the seat of a Dynahoe tractor was enormous. Triple-chinned, balanced on legs like flying buttresses, he must have weighed three hundred pounds. With the front bucket on the tractor he had been helping erect the stands for wire saws, a muscular business. He seemed almost big enough to do the job by hand. Underneath carpenter's overalls he wore a snowy T-shirt. Yellow felt gloves wagged from his rear pocket. Gray dust filmed over his catsup-colored hardhat. Seeing Jeff's camera, he drew up one pantsleg and gave a chorus-girl shake with a beefy calf. Sluggishly, yet with ponderous grace, he moved toward us over the red mud of Maple Hill Quarry through the muggy heat of June, like a diver over the ocean floor.

"It don't lack a whole lot of being hot," he said, wiping his face with a bandanna. He flashed a high-candlepower smile. His name was Lynuell Morron. "It sounds about like 'moron,' and you can make whatever you want out of that." Once the wire-saw stands were rigged in place, he would be the man to run them. He had only just come back to work after having been laid off eight months. "I sat by the fire and read the Good Book. That's the only Book for me." You could hear the capital letters in his voice.

During the winter he had also rambled around the countryside to auctions, estate sales, secondhand shops, buying and selling and swapping. "I like to buy them jumbled boxes where you don't know what's in them. Guess boxes. You might find you a wrench, say, hid down in the bottom, and sell it for more than you paid on the whole box." When he talked about the pleasures of trading, his voice took on the passion of a stockmarket wheeler-dealer. "I'll trade anything. Tools, tractors, horses, cars, chairs, crockery. I'd trade a three-legged elephant if I could get ahold of one. It's a crazy thing for a grown man to be doing, but there it is. I caught the bug off my father. He ran a country store, handled everything under the sun. There was stuff hanging from the rafters and every inch of the walls, and nobody knew where anything was except my dad. All that buying and selling got under my skin. Gave me an itch. But I didn't want to run no store, though. My dad worked from six in the morning until ten at night, six days a week, every blessed week of the year. I'd rather come out here, put in my eight hours, and go home a free man, do my trading on the weekends and when I'm off in the winter."

Morron quit school at sixteen to work in a sawmill. Then he repaired tires for a while at sixty cents an hour. He kept hanging around the quarries, asking for work, because they paid nearly three times what he had been earning in the tire shop. Eventually a ledge foreman took a look at his shoulders and put a shovel in his hands. "I was nineteen then, I'm forty-eight now, and I ain't never been out of a quarry in all those years. Things has changed a whole lot in my time. Back when I started, there was maybe sixteen hundred men in the quarrying union. Now I expect there's less than a hundred. Back then, you'd have fifty or sixty men working in one pit. It was just like ants, crawling over the rock. Now maybe you'll have eight or ten. And used to be all the lifting was done with derricks. There was twelve of them in this one quarry. But all of them's been took out, and we use that big old Tee-rex to move stone." As we talked, the huge General Motors "Terex" forklift roared back and forth across the ledge, its rubber tires, higher than a man's head, leaving canyons gouged in the mud. Cottonwood fluff drifted in the tire ruts like airy snow. It snorted past a wall on which someone had sketched in red chalk the traditional picture of the long-nosed, beady-eyed snoop peering over a fence, with the caption "Kilroy Was Here." Grunting and lurching forward, the tractor passed another wall on which someone had drawn a bare line for the fence, with the legend

> It Seems
> Kilroy
> Never
> Got This Far!

STONE COUNTRY

"There's all kind of wits in a quarry," said Morron, "including half-wits." He watched the forklift waddle out of sight with its load of stone. "Without derricks working, it don't hardly seem like a real quarry." He gave his face another swipe with the bandanna. "It's a whole lot safer today, though. No two ways about it. It was fearful dangerous in the old days. With four or five derricks swinging stone around, you had to watch out for your head every second. And with all them channelers going you had wires running every which way, and men getting shocked right and left. I grabbed a cut wire one time, and like to never did let it loose."

At Maple Hill the channelers had been supplanted by wire saws, which ran twenty-four hours a day. Morron worked all three shifts by turns, but he preferred working days. "Evenings and graveyard shift, there's only two guys here, and the time gets awful long. You mind the wires and tighten the cranks and shovel sand. But mostly you just sit." Where they would sit during cold weather and rain was a hutch that looked like a tin privy. Rusted, its door hanging awry like a broken jaw, the shed seemed too cramped to hold all of Lynuell Morron. "When evenings and midnights are sweet, though, is in the spring and fall. You build yourself a fire to keep warm, and the sky's clear and snapping with stars, and they ain't nobody looking over your shoulder to tell you what to do, you got only what's in your own head to rely on, and all there is in the whole wide world is you and a fire and the hum of that wire cutting stone."

"It takes a particular kind of guy to stick it out in a quarry," said Carl Anderson, a pantherish ledge foreman at the B. G. Hoadley Quarry in Guthrie. "You can pretty much tell right away if a guy has it in him. And it ain't just muscle. There's big old strapping guys come in here and don't last a week, and scrawny guys that last forty years."

The Men in the Trenches

So far, since moving up from Kentucky to get a job digging stone, Anderson had lasted eighteen years in the quarry, half his life. Quick in mind and body, he had worked his way up the ladder—shoveler, breaker, drill runner, wire sawyer, stonemarker, and now ledge foreman. He was no straw boss. Up he scrambled onto the ledge where the saws were grinding, down onto a freshly turned cut, over to the stack of fresh blocks, picking up idle tools, throwing an arm across the shoulders of his men, leaving whirlpools of laughter wherever he went, prowling with the agility and fierce physical presence of a great cat. A ruff of black hair spilled out from under his yellow hardhat. LET'S GO AMERICA, the hat proclaimed. Unmarried until four years ago, he still dropped by now and again to visit his bachelor taverns. "But when the girl sitting across the table from you starts to looking awfully good, and she ain't your wife, you know you better get on home pretty quick." On Sundays he golfed. His eyes, when the sunglasses came off, were gleaming chestnuts, quick and mobile, alert to the dance of men and machines. Even while at rest, leaning against a block to answer my questions, he seemed to be in motion.

"There's two things you learn when you go to working in a quarry. You're going to get tired as a dog and filthy as a hog. When you lay down at night you know where every bone and muscle is in your body. It's like farming—outside in any weather, wrestling with dirt. Half the guys here started on farms. Instead of raising corn we're raising stone."

"Do you know where the stone goes, what it's used for?" I asked.

"Not usually. I know there's some stuff from our mill gets shipped out east. New Jersey and New York, and like that. But to tell you the truth I really don't think much about what happens after the blocks leave here. I just want to get them out of the ground, clean and square."

STONE COUNTRY

In eighteen years he had never been laid off. "But most of the guys, they're off and on. Business goes way down and they get sent home. Business comes back up, and here they come again. Some guys go to working at other things where it's more steady and the money's better. But some guys just never can shake a-loose of quarrying. It gets in your blood, seems like."

A sad-eyed man, about forty, bald on top and with scraggly mutton chops on his nerve-tight jaws, sidled up to ask for a job. "I've been hunting work high and low for more'n six months," he said in a whispery voice.

Anderson looked him over. Everything about the man sagged—corners of mouth and eyes, shoulders, knees. "We're full up right now, buddy."

"I'll do anything."

"You ever work in a quarry before?"

"No, but I learn quick. And I got a good back."

Anderson checked the time by squinting at the sun. (A watch wouldn't last a month here in the wet and grit.) He grunted. "Well, leave your phone number on a paper in the office."

"I ain't got a phone."

"Then leave your address."

Later, eating lunch at his desk in the office, Anderson found the slip of paper. "What in the hell's this?"

Anderson's younger brother, Larry, one of six quarriers chewing sandwiches in front of a squealing window fan, said, "Wasn't that the sorry son of a bitch looking for a job?"

"Oh, him." Anderson wadded the paper into a ball and dropped it into the wastebasket. "He sure didn't have it, did he?"

"No, he didn't," the younger brother agreed. Less brawny than Carl, Larry Anderson had the same swiftness of hand and speech and laughter. His face was a fox's, wide at the forehead and pointed at the chin, with a sly mouth and staring dark eyes. This morning I had seen him run the drill, crane, and bulldozer, swing the hammer and sling the shovel. But his chief job was to grade and mark the quarried blocks. The stack he climbed was eighty feet high today and would rise another forty or fifty feet. "You can feel it shifting under you when it settles." In sunlight the pile of creamy stone looked like a village of whitewashed houses hugging a hill. For eight hours he billygoated up and down the stack, guiding the derrick, labeling the blocks with red spray paint. "I go home at night with miseries in the knees."

The office where the men were eating lunch was about the size of a dump truck, a two-room house moved from a farm across the road. It looked battered enough to have crossed an ocean. The clapboards had been painted white a long while ago. License plates had been nailed over holes in the floor. A barrel stove squatted in the middle of one room, ringed by clean footgear for the trip home—loafers and oxfords and carnival-colored running shoes. Coats dangled from pegs on the walls. In the other room, where the men sat on upended logs, one corner was piled with spare hickory handles for the stone hammers. The wall separating the two rooms was shaggy with the peeling remains of flowered wallpaper. A calendar, with X's through the finished days, showed a toothy woman in a grandmotherly swimsuit holding a wrench. The only thing to read was *Bowhunter Magazine*.

They talked about hunting, about cars, money, their new girls, their old ladies. Three or four times they spoke casually of niggers, using the ugly word unthinkingly, without venom, as if it were a neutral noun. Had they ever known any blacks to work in the quarries? No, never, they agreed. And why was that? A man lowered his can of Mountain Dew and said, "I guess they don't like our kind of hard work." Then how about women? The notion made them hoot with laughter. How would a bunch of bulls keep their mind on business with cows rubbing up against them? they demanded.

"Depends on the business," said Mountain Dew.

"If we had women out here," Carl Anderson said, "that's the last lick of work I'd get out of these guys."

"The girls might get a few licks."

"They'd wear out our tools in a hurry."

"Dull our drills."

"Lay off the women," said Anderson. He seemed uneasy about the rowdy drift of the talk. However rough the surroundings, he thought of quarrying as an ancient and honorable craft. Leaning toward the fan, his hair fluttering out as if electrified, he said, "What we need more than women is a shower." Now you're talking, everybody agreed. "The coal miners has all got showers. Why, when they leave the pit they look like durn office workers, they're so clean, and we go home like a bunch of hogs that's been rolling in the mud. Why don't we rate like the miners?"

Another lunch hour, this time at Reed Quarry north of Bloomington, five men were sitting on the bed of a pickup in the shade of an engine house. They traded their food around. They were built like Sumo wrestlers, thick in legs and neck and even thicker in between, their girth swollen by heavy lifting and heavy eating. Their old ladies were after them to cut back on food.

One heavyweight used to bring four sandwiches, an apple, a banana, and half a gallon of milk, finish it all before lunch, then go home at midday for more grub; now his wife sent a can of tuna fish. "Only a young buck can eat like that and still keep working," he said. "Nowadays I got to keep my stomach light, or I can't move around down in the hole. As it is, when I go home after climbing ladders and standing on rock all the day long, my feet feel like they're covered with boils."

A second man, the biggest of the lot, carried the same lunch every day, year in and year out—hotdogs and applesauce. There was a musing, inward look on his face as the other men kidded him. They had to grab him by the shoulders and shout into his ear in order to penetrate his deafness. "Ain't that true, you old ox? Weenies and applesauce?" He stared at his boots and grinned shyly.

Eating and teasing, they leaned against one another, flung arms around necks, stuck elbows into ribs, prodded and patted. This gruff physical intimacy came from working together in tight places amid the racket of machinery. They were used to touching while they talked. They singled out a goat for today's ribbing, a red-faced fireplug of a man whom they called Buttermilk. "If you ever see that guy working, take a picture, because we all want to know about it." Three toothless men showed off their gums. "As soon as we come into the money we'll buy us some new clappers, won't we? Or maybe we'll go down to the funeral home and pick us up a pair. They say they've got a whole basket of leftovers at the funeral home. We'll get us some with gold teeth."

After they finished eating, crows landed in the pickup and gleaned their crumbs.

The man who watched over them when they returned to work was Steve Reed. About thirty, quiet, he possessed the physique of a gymnast and the unmarked handsome face of a talk-show host. With such a face he could have observed the world, if he chose, from above the rim of a white collar. But the shirt he wore was brown, baggy, and splotched with sweat. "I'm the fifth generation in the Reed family to work in stone. When I was little, I didn't think I'd like it. Dad always made it sound like the hardest way on earth to earn a living. And it's pretty close to that, all right. But in high school, I came out here during the summers and worked as a breaker and drill-runner, and I got a taste for it." After studying business at Indiana University, he had come to work in his father's pit full time, grading the stone, deciding where to strip and cut and stack, orchestrating the motions of forty men.

Like most quarrymen, he had a winter passion. "I trap foxes. Last year I caught forty-five, adding the gray and red together, and that's the most for anybody in Monroe or Owen County. I sell the pelts, of course. But I don't trap for the money. People think you have to hate an animal in order to kill it. I don't hate foxes. I love them. I'm fascinated by them. I've studied them for years, and read everything about them I could lay my hands on. To catch them, you've got to think like a fox, and that's hard to do." When he found one in his trap, he would finish it off with a quick blow to the head. "But I've got to admit it's harder for me to kill them every year. I'm getting soft on them. I wish the Department of Natural Resources would shorten the season, so they don't get killed off." The foxes were being snuffed out not only by trappers but also by coyotes. He pronounced the word "ky-oats." "They're all around here, and they're bigger and stronger and meaner than foxes. Some people call them brush wolves. They'll eat anything and do anything to survive, and that includes killing foxes on sight. They'll also kill sheep and calves. I've had farmers tell me about finding a calf with its hind leg chewed right off, and the calf still alive. They're hard as hell to catch. I worked at it all last year and trapped five. They're tough hombres, I'm telling you. Real survivors."

Down in the pit, a man who had been gathering scrap stone into piles stooped to wash his hands in a puddle on the quarry floor. The opening was a hundred feet square, a small bite from the flank of a ridge. "There's enough good stone in that ridge to last me my lifetime," Reed said. He planned to keep at it; but he might be the last Reed to do so. One of his father's brothers had already given up quarrying to sell golf carts. "I don't blame him. Why wade in mud and snow to dig rock out of the ground, when you can earn twice as much sitting in an air-conditioned show room?" Five generations back, the Reeds had been masons and cutters in Chester, England, a city ruled over by a stone castle, inspired by a stone cathedral, and encircled by a Roman wall of stone. Steve Reed had a newborn son, the first member of the sixth generation. Did he want the boy to become a stone man? "I don't know," he answered, his eyes following the glide of a block as a derrick landed it on the quarry lip. "It's my life. But it's a hard and risky business, and I just can't say if it's what I'd want for my son."

8 Cutting

We followed a flatbed truck hauling a pair of biscuit-colored blocks from Reed Quarry to Woolery Stone Mill, southwest of Bloomington. Viewed from a rise as we approached, the mill looked big enough to serve as a hangar for blimps. The pickups in the parking lot were all aimed toward the highway, ready for a quick escape at the end of shift. Jeff and I entered through a door that was flanked by hearty stands of joe-pye weed, their fuzzy purple heads growing taller than my fuzzy brown one.

Jack Rogers, third in a line of fathers and sons to manage the place, showed us around. When he had started in the mill as a highschooler during World War II, the chief product was not limestone but parts for army trucks. "The Depression nearly wiped us out; but Hitler kept a lot of operations going. I know some other mills that stayed open by making hulls for Sherman tanks and aluminum cylinder heads for bombers." Now in his fifties, portly, with thinning hair and metal-framed glasses, Rogers was soft voiced and melancholy and a little stooped, like a discouraged priest. And the mill he showed us through was like a cathedral: vaulted ceiling held up by steel girders, light streaming through high clerestory windows, a nave so long its farther end was vague with dust and distance, and everywhere the cool presence of stone. Leaning in a corner, to complete the effect, was the plaster model of a saint.

"Back in the heyday of limestone, right after the war," said Rogers, "they just kept adding on to this place, building it bigger and bigger. Now, with business so far down, we don't need all the room." So many pews, and a dwindling congregation.

Everything was powdered white from stone dust, as if a snow had fallen. Light slanting through motes in the air appeared substantial, like pillars of glass. I worried aloud about the lungs of the workers. (Talking with men in various mills, Jeff and I had been told about coughing miseries that sounded very much like silicosis.)

"I've never heard that it causes any problems," said Rogers. "They say it passes right on through your system. Most of the men live to a good old age." From the moment we arrived he had been smoking Kools. Drawing the current cigarette from his mouth, he gave it a squint and said, "This is the stuff that will ruin your lungs."

One end of the mill stood open, to allow movement of blocks from the stacking yard inside to the saws. Enough sunlight leaked through gaps in the steel curtain walls to encourage saplings and weeds to root in the limy dirt of the floor. The I-beams supporting the roof were braided with ivy. Pigeons and starlings swooped in and out through holes in the windows where panes were missing, and butterflies wafted in to settle on the golden buttons of tansy. Outdoors merged with indoors. Near each work station there was a potbellied stove, idle now in August but fierce with coke fires in winter. As in most factories, there was a sound of motors keening and belts rumbling. But the ground-note of metal rasping against stone was peculiar to such mills, a watery grating as of creeks over gravel. "I don't even hear it anymore," said Rogers, "it goes down so deep in me." The smell was also distinctive, calling to mind tidal beaches, damp sand, river deltas.

It was a male preserve. Only men cut stone, as only men quarry it. In some mills, even the clerks and secretaries are male. No one, including Rogers, could explain to me why this is so. "I guess it's just one of those old-fashioned things," he said. "I've never heard of a woman working in a mill. Never even thought about it before." It's muscular work, but there are women who could handle the machinery as well as the men do. Perhaps those who wrestle with stone imagine it to be female, the flesh of Mother Earth. Or perhaps men keep this work to themselves simply because it is dangerous and dirty. Whatever the reason, the lone female working here at Woolery was the receptionist, and otherwise women appeared only in pinups tacked to the wall, as passive as the waiting stones. (I should also say that we never saw another pinup in any of the quarries or mills we visited.)

"Don't mind the artwork," said Rogers.

At every step in our tour through the mill, he brushed his fingers over the stone, a stroke of familiarity and affection, as a farmer goes among the beasts in his barn. He touched the pieces of various jobs, identifying them for us: balusters and stair treads for the mansion of a Saudi Arabian prince in Virginia, panels for a Texas bank, foundation blocks for a Philadelphia synagogue, gargoyles and filigree for a college tower in Missouri, an abstract sculpture for an Indiana restaurant. He also touched the rejected stones. Like a farmer reading the condition of his cows, he could read the faults and strengths of each piece with a quick glance. "These black jagged lines are called crow's-feet," he said, fingering a slab in the waste bin. "They'll ruin a stone. And so will these hairline cracks here from cross-bedding, and this change in grain from coarse to fine. The coarsest stuff we call gothic, like this here, with fossils in it as big as dimes, and little air pockets to give it texture. That's my favorite limestone, but you can't sell it to people. They take one look at those shells and get nervous. But it's just as durable as the finest-grained stones."

For my money, I told him, the bigger the fossils the better. I'd love to have a house molded entirely of fossils, a shell of shells, a skin of bones.

"The only trouble with coarse stone," said Rogers, "is that it plucks on the planers and leaves a rough surface."

The planers and lathes and saws were built on a scale for giants. The normal-sized men who ran them were mostly sons and grandsons of stone workers. Their ancestors moved into this region from Tennessee and North Carolina and Kentucky, or immigrated here from Italy, from Austria, Hungary, Yugoslavia, Greece. In the early days, owners sometimes imported foreign workers to break up strikes. The local men resisted fiercely, sometimes with fists and clubs. But many of the immigrants stayed, joined the union, and went on strike in their own good time. "It seems like people in the stone industry never believe they've got a fair contract unless they go out," Rogers lamented.

Organizers from the American Federation of Labor entered the district about 1900, signing the men into unions with mouth-filling titles such as the International Association of Marble, Slate, and Stone Polishers, Rubbers, and Sawyers, Tile and Marble Setters Helpers and Terrazzo Workers Helpers; the International Brotherhood of Blacksmiths, Drop Forgers, and Helpers; and outfits with simpler titles, such as the Journeyman Stone Cutters Association, the International Union of Stationary Steam Engineers, and the Quarrymen's Protective Union. They mounted their first strike in 1903.

"How often do they go out now?" I asked.

"Oh, about every three years or so," Rogers answered.

"How often do you negotiate contracts?"

He gave a priest's muffled laugh. "Every three years."

Descended from men who picked up and moved when things didn't suit them, workers in the limestone country have always had itchy feet. In the 1920s, a typical mill lost three-quarters of its men every year. One mill had a turnover rate of 250 percent. Some of this moving about was the result of piratical raids by rival owners, but most of it was due—as much of the shifting by settlers from clearing to clearing along the frontier was due—to sheer restlessness, a hankering for greener grass. Journeymen stonecutters in particular have lived up to their name and journeyed all over the country—Chicago, Atlanta, Boston, Washington, Houston—wherever the pace of building is feverish and the wages are high.

"Turnover's nowhere near what it used to be," Rogers said. "The pay and benefits are about the same everywhere in the belt. Besides, there just aren't as many jobs waiting for the drifters."

Work in the mills has always been precarious, boom and bust—in recent years less often boom than bust, with the building industry sickly much of the time and with steel and glass and concrete replacing stone. Even in fat times the work was never secure. Prosperity enabled the owners to buy improved machines, stronger ones and faster, and each mechanical advance reduced the need for men. Between 1907 and 1915, when pneumatic tools replaced the old mallet and chisel, about half the region's stone cutters were thrown out

CUTTING

of work. There were bitter strikes, but technology won out over handcraft.

The Woolery men glanced up and snapped jokes at us as we passed by, their arms decorated with tattoos, their noses flattened like the noses of boxers.

Rogers excused himself to return to the front office, where he would punch the keys on his calculator to figure bids for future jobs, and where, at lunchtime, the draftsmen and supervisors would play passionate games of euchre.

Jeff and I sauntered around on our own, tracing the path the limestone takes on its journey through the mill. A typical quarry block, when it arrives at the stacking yard, is the size of a panel truck and weighs between ten and twenty-five tons. By the time it is loaded on another flatbed at the far end of the mill, a finished piece, weighing perhaps a few hundred pounds, will have been cut to fit snugly in a building. Every piece is accompanied by a schematic drawing called a job ticket, which draftsmen make up from architects' blueprints, and in its course from saws to planers to cutting benches it gradually takes on the ideal shape described on the ticket—a section of an arch, it might be, or a wall panel, a baluster, a column. Piece by piece, like the thousand bits of a jigsaw puzzle, entire buildings leave the mill on trucks.

Electric cranes rolling overhead on elevated rails lift the quarry block from its resting place in the yard, carry it inside, and set it down on a miniature flatcar. The flatcar trundles the block under a gang saw—a set of parallel blades fitted into rocker arms—and here the block is carved into slabs like a Thanksgiving turkey. Before the introduction of gang saws in the 1850s, the cutting of stone had changed little since the time of the pyramids. Masons used to crack boulders by heating them in a fire and then dousing them with cold water. For cuts more regular than fire was likely to produce, old-time masons pecked at stone with hammers and chisels, or bored a string of holes with hand drills and split the stone with wedges, or tamped the holes full of black powder and blew the stone apart. Harder steels developed in the nineteenth century made it possible to fashion crosscut saws for slicing stone, much like the ones lumberjacks once used in felling timber. These back-breakers were introduced into the limestone country by an Englishman named John Matthews, who had learned to cut stone while helping to build the Houses of Parliament. Gang saws evolved quickly from the crosscuts, replacing the single blade with six or eight blades in tandem, replacing the two stout men with a steam engine and, later, with an electric motor. At first the saws were toothless; they sliced through stone by grinding sand into the cut. (On our arrival at Woolery, we had passed a heap of worn blades rusting in the high grass.) Later, the blades were rigged with teeth, some tipped with diamonds, others with carbide steel.

One of those who watched over the gang saws at Woolery was Ernest Hattabaugh, a small, dark-featured man with a gray wave in his slicked-back hair and a going-out-on-Saturday-night smile. The skin of his face was gullied and cracked like a rain-starved field, and the flesh was wound as tightly on his frame as wire on the armature of a motor. Quitting school at fifteen, he had gone to work trimming trees, building railroads, sawing lumber—"just whatever I happened onto"—before getting his first job in a mill twenty-nine years ago. Like many stone men he began in the yard, breaking up rough-backs—the outermost slabs left over when a block has been sliced up in the gang saw. "They loaded them rough-backs into a gondola car, and you had to climb all over them like a mountain goat and bust them up with a twelve-pound hammer until they wasn't any pieces bigger than would fit through the mouth of that gondola. The train hauled off the broke-up rock to a crusher, and they made it into ballast for the railroads or lime dust for farmers." Hattabaugh spent a decade hammering rough-backs. "I tell you, it like to killed me. It about wore me down to a puddle of grease."

Eighteen years ago, he had been released from the purgatory of the gondolas to become a sawyer, and there he had remained ever since. "It's work I like. So long as I keep my eyes open, they ain't many surprises. I just watch the blades and listen to them beat to make sure they're rocking smooth. Every fifteen minutes or so I got to go have a look through the cuts, to see they're straight. If I can't see daylight, the blade's bending. And there's maybe a hundred and fifty points I got to grease every week. Once like a fool I slipped down between the beater arm and the block, and got my ass pounded all black-and-blue. But by and large I ain't been hurt."

From the gangs, freshly cut slabs are delivered by traveling cranes either to the planers, for finish work, or to the diamond-tipped circular saws for additional ripping. The dean of diamond sawyers at Woolery was Marvin Brinson, a sixty-year-old who looked forty—who looked, in fact, not so much young as ageless, with the wry, smooth, philosophical face of a meditating yogi. What he meditated, both at work and at home, was the Bible. A "Good News" paperback edition of the New Testament lay facedown on his stool, with a pair of reading glasses folded across the spine. Political cartoons and handwritten quotations from his reading—but no inviting girlies—papered the wall behind his seat. A sign declared: CAUTION—HEARING DANGER ZONE.

With a wrench the size of a baseball bat, he cranked a small flatcar into position beneath the arm of his saw. When the slab on the car was lined up to his satisfaction, he turned on a spray of water for cooling and set the man-high circular blade in motion. He checked to see how hard the saw was working by reading a meter that showed what amperage the motor was pulling. As the diamond teeth whined through the slab, we talked in hand signals and in bursts of words shouted into one another's ears.

"How long have you been a sawyer?" I yelled.

"Thirty-five years."

"How long was your apprenticeship?"

"I didn't have one. The foreman sat me down here, pointed at the switches, and walked away." He tilted a smile at us. "I could drag somebody in off the street and teach them to run this thing in half an hour. Why, I could even teach you two guys."

"Do you ever get used to the noise?" Jeff shouted.

"What noise?" His smile tilted the other way. "Really, all that bothers me is the water. It makes my joints ache. Most of the old-timers get the arthritis pretty bad."

"You aren't an old-timer, after thirty-five years?"

"Heck, no. I'm a spring chicken."

The saw roared to the end of its cut, and the relative hush that followed came like the sweet relief after headache. Brinson cranked the flatcar into a new position and let the blade rip once again.

He pulled out his wallet to show us snapshots of his prize vehicles—a Corvette, a dune buggy, a three-wheeled Volkswagen Beetle, a souped-up Model-T—each one painted or sculpted or otherwise transformed in ways bizarre enough to make you doubt you would ever see another one like it.

"Do you customize them?" I said.

"No, I buy them like that. I don't work on cars. I just collect them."

"Do you drive them?"

"Some of them. Right now I'm carrying licenses on eight vehicles. And insurance! The insurance is just awful."

"What do you drive to work?"

"A beat-up old pickup. You don't think I'd risk one of my fancy jobs around these cross-eyed cowboys, do you?"

A planerman slouched by on his way outside to have a piss in the weeds, and he and Brinson kidded one another about who was sweating harder.

"Cats been scratching your stone?" Brinson hollered at him.

"My stone's smooth as glass," the other man yelled back.

"What do you mean, cats?" I asked, while the planerman was relieving himself in the joe-pye weed.

Brinson crooked a finger and led us to a corner where an upended five-gallon bucket rested on the dirt floor. He lifted the chunk of stone that had been weighting it down, tipped the bucket, and revealed a loaf of bread. "I put out food for the wild cats and dogs that live around the mill, and for raccoons, squirrels, birds, whatever's hungry."

Zipping his fly, the planerman shuffled past again. "Did Marvin tell you he comes in weekends and holidays to feed his blessed animals?"

"And why not? Their stomachs don't know weekends from weekdays, do they?" Brinson lowered the bucket over his bread. "If I didn't look after the animals," he said to the planerman, "where would you guys be?" And then to us, "These guys, you know, if there's any scratches on the stone after it's left the planers, they blame it on the cats that work the nightshift."

Planers have been around for a hundred years, and have changed little in that time. Huge, grease blackened, held together by bolts the diameter of beer cans and nuts fatter than fists, they look like torture machines from the dawn of industry. The stone is wedged in place onto a rolling bed that is about as wide as a double mattress and two or three times as long. The planerman fixes his cutting tool onto an upright stanchion, adjusts the depth of cut by twirling spoked wheels and tightening screws, and then sets the bed in motion. Dragged against the tool, the stone peels and powders away like soap from a fingernail. Pass after pass, increasing the depth of cut each time, he scrapes the stone into the desired shape.

What tool the planerman chooses depends upon the profile he wants—wide, flat chisels for smoothing surfaces, curvy bits for moldings, rounded shovels for fluting. In scarred wooden bins behind the planers at Woolery there were hundreds of tools, in more shapes than a beachful of sunbathers. They had been custom-made out of tungsten carbide steel by the mill's own blacksmith, one of the vanishing trades still kept alive in limestone country. Smithies forge and sharpen not only the planer blades but nearly everything else that's made of metal. Craftsmen possessed of ancient skills, they tend and mend antique machines.

In the blacksmith's shop at Woolery, two mechanics and an electrician wearing oil-stained blue shirts sat in lawn chairs talking about cats.

"Old Tom, poor Tom."

"What's the trouble with Tom?"

"He's never been the same since you stuck his head in that boot and cut off his nuts."

"But there ain't so many kittens underfoot anymore, is there?"

The workbench was loaded down with items in need of repair—yokes, chains, cogwheels the size of car tires, motors the size of pigs—but the talkers seemed to feel no urge to rise from their lawn chairs.

Cutting

"It's no wonder he looks so down and mopey, with his nuts cut off."

"You rather have bushels of kittens?"

The mouth of the forge was filled with ashes. Sledgehammer resting on the anvil, enormous jaws of vises gaping, drill presses, grinders, lathes, arc welders with bottled oxygen—everything idle, and everything black—fire-black, grease-black—like Vulcan's workshop in the underworld.

As I stood there enjoying the heavy look of the tools, one of the mechanics leaned back, hands clasped behind his head, to watch me through slitted eyes. My stillness must have seemed mysterious to him, for at the moment I was not talking, not writing on my notepad, doing nothing at all except gazing. "You working hard, are you?" he said.

"Flat out," I said.

"You really interested in all this old stuff?"

"I sure am."

The mechanic took off his Beechnut Chewing Tobacco hat, raked black-edged fingers through his hair, and said, "The way I look at it is, if you don't know nothing about limestone, it's interesting as all get out. But if you growed up around the mills and quarries like I did, it ain't."

Jeff, meanwhile, was rigging his camera, lengthening the telescoped legs, screwing parts together. Here was a labor the men in the blacksmith shop could understand.

"You must be studying to become a foreman," the mechanic said to me, "because all you do is stand around while your buddy here does the work."

Jeff agreed.

Just then a real foreman entered the shop to announce that a motor had burned out on one of the jointers. The lawn-chair loungers exchanged disgruntled glances, no one anxious to move; then the one who had castrated the cat rose with a grunt and went to have a look at the trouble.

Another diamond-tipped circular saw like Marvin Brinson's, only smaller, the jointer is where stone goes after the planerman finishes with it. Sliding past the saw on fat rollers, each piece is cut to length. From here, a traveling crane lifts the stone in a pair of thick woven belts, dandles it in the air, and sets it down on stout wooden benches near the truck-loading ramp. The men who loop and unloop the belts, like those who attach the doghooks to blocks in the quarry, are called hookers—tail hooker and head hooker. With nearly imperceptible nods and finger flicks, they signal to the traveler-runner, who sits in a cage high overhead, tugging levers and pushing pedals, jockeying the crane back and forth on its elevated rails.

When they leave the jointer some pieces are ready for loading into the beds of low-walled trailers. Until the 1930s, everything rode by rail, both into and out of the mills. Today almost everything goes by truck. In the old days the stones were packed in wood shavings: occasionally sparks from the locomotive would set a load afire and blacken the stone. Today the two spry men who pack the trailers—still called car-blockers from the days when they loaded railroad cars—place strips of fiberboard between the layers of stone and stuff fistfuls of sisal into every gap.

Other pieces, when they leave the jointer, must be drilled or slotted for the anchors that will hold them in place on a building. Still others must be given shapes more subtle than anything the machines can produce—and this final flair of subtleness is the work of the aristocrats of the stone trade, the cutters and carvers. Working on blocks that have been sawed and planed to the rough dimensions of the job, they peel away stone with air-powered chisels, fashioning gargoyles and ivy, sculpting the statues of a bishop or a rhinoceros, inscribing angles and letters and curves. The sound they make is of a hundred ferocious woodpeckers. Dust disguises their clothes and turns them into ghosts. If they were to put down their air hoses and pick up mallets and chisels, they could easily be at work on the Louvre, the Alhambra, or the Parthenon. It is an ancient art they practice, hand and eye coaxing shapes from rock.

It is also a vanishing art, because modern architecture—obsessed with clean lines and contemptuous of ornament—has largely dispensed with the ornate carving that was essential to the revived classical architecture of fifty and a hundred years ago. "All they put up nowadays is crackerboxes," an old stone cutter complained to me, "just flat walls without cornices or pediments or tracery, without a bit of art or shadow to them. They look about like parking lots stood up on end and window holes punched through them. With all the new buildings so plain, where does that leave a stone cutter?"

At Woolery the few remaining cutters, who had learned their trade working on groined ceilings and rose windows, were chiseling initials into the gateposts for a Georgia condominium. One of them pushed his dusty spectacles onto his forehead, rubbed his eyes, and said, "This here's piddling work. You could take a boy apprentice six months on the job and he could do this. But there's precious little of the really good work ever comes in anymore." The creases about his eyes were white from powdered fossils. He was himself the next thing to a fossil, one of the few surviving members of an endangered species.

On our way out through the front office, Jack Rogers gave me a rectangular slab of stone about the size of a bread slice. "You said you liked the gothic grade," he told me. "Well, here's a piece to take home."

STONE COUNTRY

While Jeff packed his gear in the trunk, I lay my present on the hood and bent down to study it. Spirals, webs, curlicues, ringlets: it was a fruitcake of fossils. It was all the shapes of a galaxy shrunken to fit my hand. Comets, meteors. As many creatures had gone to make up this handful of stone as there are stars in the Milky Way. The rectangle became a door: if I stared at it long enough, it would open. If I stared until the mammoth scale of the ordinary let go its grip on my mind, I would become small enough to swim through these microscopic nebulae.

I was still bending over my slice of stone when a klaxon sounded the end of shift. Men trooped past me with lunch pails clunking against their thighs.

"You okay, buddy?" said a nine-fingered sawyer.

"Fine," I answered, just fine, snagged there by the gravity of a hand-sized galaxy.

9 Three Carvers

"In Bedford, when I was a boy, there were ordinary families and there were stone families. I came from a stone family. My father ran a lathe. My uncles were carvers. Every morning I'd sit down to breakfast and all the talk was stone, stone, stone. At supper it was the same. So naturally, I went into the mills and learned to carve."

The man telling this to Jeff and me was Jack Kendall, who had stuck at carving from the age of fourteen until the age of seventy-eight. Now he was eighty-two, but his years showed only in the forward slump of his head and shoulders when he rose to shake hands with us, and in the sharp creases beside his mouth. His forearms were tattooed with a star, a girl in a sunbonnet, and a cross wrapped in flowing drapery. The veins stood out across his muscles like the negative images of rivers.

We were sitting on the front porch of his stone-and-block house in Bloomington. Across the road, in what had recently been a meadow, was a cluster of plywood apartments that looked as hasty and temporary as a clump of mushrooms. While we talked, three grade-school boys danced on one of the raw sundecks to the boom-ba-ba-booma blare of a radio. Kendall took no notice. His hearing was fine, and so was his vision. The eyes he turned on us were the soft gray of old milkweed fluff, but they saw near and far without benefit of spectacles. He was oblivious to the dancers, the radio, the jacked-up cars booming by on North Maple, to everything except his memories of carving.

"I probably know as much about stone as anybody in America, because I've cut every kind there is. Lavas, tufas, sandstones, granite, marble, limestones—everything. I can look at a piece of rock and tell you just about exactly where it's from. You might see sandstone, but I see Pennsylvania sandstone or Ohio sandstone or any of a dozen other types. Every kind of rock has its own personality for me."

When his hands were not occupied with lighting and stubbing out Lucky Strikes, they sketched the shapes of his words in the air. His fingers were as narrow and quick as a guitarist's; on one of them he wore the heavy gold ring of a thirty-second-degree Mason. In blue knit shirt with white striping, gray flannel trousers, and black dress shoes looped with chain, he looked quite dapper. The carvers he learned from sometimes wore fedoras, bow ties, and white shirts. "They knew they were somebody. You could *see* they were special in the way they carried themselves." The names these carving masters wore had come fresh from England—Adams, Andrews, Elgar—from Germany—Meissner, Bruner, Ditzenburger—and above all from Italy—Donato, Calucci, Mazzola, Solomito, Liva. "To me, when I was a boy, a carver was about the grandest sort of man you could become." Talking about these heroes, he sounded like Mark Twain talking about the glorious river pilots.

Kendall made his start at fourteen during the Great War sculpting tree trunks and doughboys out of limestone. "The trees really *looked* like trees, with stone ferns growing around the base of them, ivy climbing up the side, and growth rings in the nub ends of branches. You could close your eyes and run your hand over the bark and it felt just like an old stump. Some people put them in their yards. But mostly they were for grave markers. A railroadman who had five children might die, say, and they'd order him a tree trunk with five limbs coming out of it and a locomotive running over the roots. The doughboys were shipped all over the country for war memorials." His older brother swallowed mustard gas in that first no-holds-barred war, and never recovered.

In the twenties Kendall worked on the Tribune Tower, some of the time at the Tribune Mill in Bloomington, some of the time on the building site in Chicago, carving rosettes, grape clusters, acres of gothic fretwork. From there he went to Detroit, New York, San Francisco, Washington, New Orleans, all over the country, a man in demand. "I was a little wild in those days. Full of wine. When I went up to Montreal for a spell, I'd work all the day and then go tobogganing or skating half the night." Knocking about, he carved Indians, Uncle Sams, birds, squirrels, assemblies of saints, gigantic statues of the Virgin Mary, griffins, gargoyles, roses, and rose windows. He often worked from plaster models, sometimes from drawings or photographs, occasionally from casts of ancient sculptures. (Marshall Field III, reigning prince of the mercantile family, sent him to the Yucatan Peninsula to make copies of Mayan carvings for the Chicago World's Fair of 1933.) "Usually a sculptor provided the original, and I copied it, larger or smaller, whatever he wanted. He was an artist, making things up; I was a craftsman, carving into stone just whatever he'd imagined." There were highfalutin sculptors and down-to-earth ones. "You take in New Orleans, I was doing the carvings for Audubon Park with Charles Dodd—pretty famous back then—and he'd say to me, 'Let's knock off and get us some inspiration.' And we'd go drink some popskull Cuban rum, elbow to elbow. This was Prohibition time, remember, and the inspiration got a little strong."

Kendall's hair, still thick, had faded to the color of wet sand by the time we met him, but it used to be the color of fresh blood. People called him Red, and he had a temper to go with the nickname. "I've got some marks on me to show for it. Once I was on a job over in Columbus, and we were drinking some rotgut in a speakeasy. The room was dark, and one of the guys I was with wanted to have a look at another guy's girl. So he lit a match and held it up to her face. And that started the fight. In a minute we were all outside in a wet plowed field, up to our eyeballs in mud. And some guy laid my head open with a tire iron." He showed us the pale scar. "It used to be, you could get mad over something and have a fight and there wouldn't be any assault-and-battery charges. You'd say to a guy, 'I'll meet you down by the clock,' and after work you'd have a go-round, dust yourselves off, and that would be that. Nowadays, people just bottle everything up, stew and burn, and before you know it, somebody reaches for a gun."

Carvers have always been itinerant, owning their tools and traveling wherever whim or wages lead, yet Kendall was remarkably footloose even for this vagabond fraternity. "I moved because I liked moving, and also because people knew I could put a building together. If they had a problem getting the thing to go up, getting it to *work*, they'd call me, and off I'd go." Whenever he stayed long enough in one place, he signed up for night classes—math, architecture, engineering, lettering, drawing. "I had quit school at fourteen, and I aimed to make up for what I'd missed. I didn't want carving to be the only thing I knew how to do, so I studied enough to become a building engineer."

Three Carvers

From the Depression onward, he did less and less carving, more and more engineering. At the outbreak of World War II he was designing machine shops for Du Pont. He rode the first convoy into Pearl Harbor after the bombing, to help oversee repair of the aircraft carriers. Back in civilian life, he supervised the raising of churches, schools, banks, dormitories, office towers, anything made of stone. "But the thing I'm proudest of is working on the National Cathedral. I helped lay out the vaulted ceilings—very complicated, nothing so complicated in stone since the Middle Ages. There was a certain atmosphere there. When you walked inside it was like going into Carlsbad Caverns, except you knew this was something men had made. The best stone men from all the world were gathered there, and I was proud to be among them. There was a sculptor from Greece, an architect from Italy. There was an old stonesetter from England who'd worked on cathedrals his whole life. And there I was, working alongside him."

Of course he had kept doing a certain amount of carving—touching up the two-story-high Rock of Gibraltar on Chicago's Prudential Building, for example—right up until retirement. In the four years since retiring he had filled his yard and the yards of his children with ornaments. He showed us through his sculpture garden, past limestone benches, urns, lanterns, a birdbath resting on a spiral base like the twisted stem of a wine glass, planters carved in the shape of wooden buckets, a fountain encrusted with otherworldly fish and flowers. The centerpiece was a limestone castle that rose from a bed of petunias, pansies, daisies, marigolds, and moss pinks. Covering roughly the area of an ample kitchen table, elaborately detailed with gates, stairs, windows, turrets, and bridges, the castle looked as though it had tumbled out of a fairy tale. It was occupied by chipmunks. "I like to tell the rich boys from the east side of Bloomington," said Kendall, "that over here on the west side even our chipmunks live in stone castles." Proud, he stood there with one hand resting on a tower, his shoulders hunched forward a little from age, but still looming over his castle.

When Jack Kendall was a boy, there were forty or fifty mills pouring out fancy-cut stone, and each mill kept anywhere from a handful to several dozen carvers busy. Today there is only one full-time carver in the region—Henry Morris, who studied under Kendall. "The most important thing for a carver," Kendall told us, "is to be able to *see* the completed work before you make the first cut. Some of that a person can learn; but most of it's a pure gift. Henry has that gift."

In Ellettsville, at the Bybee Stone Company mill, a cube of stone the size of a washing machine rested on a spraddle-legged bench. Eventually it would become a Corinthian capital for the Iowa Statehouse, a thicket of acanthus leaves, whorled volutes, and clustered fruits; but now, fresh from planer and diamond saw, it was perfectly blank. Henry Morris circled it, pencil in hand, like a fencer looking for an opening. Tan, white haired, sinewy, sharp jawed, he might have been taken for a tennis coach. Near sixty, his face showed few lines, not even sun-squints around the eyes. His ballcap read HANK and his brass belt buckle spelled out HENRY. Circling the stone, he kept tilting his head to get the best view through his bifocals. From time to time he consulted a photograph that lay on another bench beside his toolbox. It showed a bleary image of the eroded sandstone capital he was supposed to reproduce.

"It ain't your true Corinthian cap," Morris said, pointing out irregularities in the photograph. "It was just somebody's *idea* of a Corinthian, back a hundred years ago. But I'll carve mine to match what those old guys did."

Eventually he would carve sixteen, enough to renew every col-

Three Carvers

umn on the Iowa Statehouse. Today, with Jeff and me looking on, he was roughing out the first one, the pattern for all that would follow. Gazing from the photograph to the blank stone, he saw in three dimensions, imagined level beneath level. On the creamy surface he penciled neat X's to locate the high spots, then he rummaged for the proper carving tool, slipped it into the pneumatic hammer, and started peeling stone. Beneath his hands the rock looked butter-soft.

"You take it down awfully fast," I said.

"I know this stone. I know what it'll do. Now marble—you got to chip that in little bitty bits, and even then you can't be sure what it's going to do on you. But limestone, I can go right through it, because I know how it cuts."

The rattle of the hammer was so loud that we could talk only when Morris paused in his work. He paused often. Unlike the other men at Bybee's, Morris was a free lance, responsible to no one. He would bid on jobs and then arrange to do his carving in one mill or another, at whatever pace he chose. The sooner he finished a job, of course, the more money he cleared. But he could think of quite a few things more important than money, and one of them was enjoying his work. "If you push, push, push all the time, where's the pleasure in that?" The day before, he had knocked off at noon to go buy a new truck. (Air-conditioning, power windows, velour upholstery, carpeting up the walls and across the ceiling, stereo with four speakers—a golden calf on wheels. "Everything but a beer machine, and I'll have to get me one of those.") Today he would stick around for eight hours, but he kept pausing to study the photograph, to take a sip of cold coffee from a thermos, to light an Old Gold, or to nip a chaw of Levi Garrett Plug tobacco. The stone dust at his feet was pockmarked with brown stains.

His wooden toolbox, snugged together by steel bands, was piled deep with tungsten carbide gouges, chisels, rippers, splitters, liners, roughing-out bits, over a hundred tools altogether, one for every imaginable cut. Loaded, the box weighed nearly as much as Morris, and more than he had any intention of lifting. Whenever he wanted it moved, he signaled for the hookers and traveling crane. At night he left it padlocked. Some of the tools he had made, but most he had bought second- and thirdhand, and these were engraved with the names of dead carvers—Moise, Baxter, Emery, Elgar.

"I worked with some of the old boys, including Dugan Elgar. His real name was Harold, but everybody called him Dugan, which was the brand name of this patching compound for hiding the mistakes you'd made in a stone. We all used the stuff, but he got stuck with the nickname. It started out as a joke, I guess, only he was such a good carver he ended up making it sound like an honor." (If you can't take a joke or give one, you won't last long in the mills. Humor is the grease for getting by. In the front office at Woolery we had seen one of Elgar's whimsies, a statue of a smirky frog lounging against a mushroom and smoking a cigar.) "Dugan was a great one, no question, but I swear he had the damnedest nose for publicity. I worked on a lot of jobs with him, and I'd do most of the carving, and then Dugan would show up with reporters and photographers and put on the last few strokes while the flashbulbs popped." In 1968, Elgar carved his own tombstone, a copy of Michelangelo's *Pietà*—Mary grieving over the crucified Jesus—and placed it in the Clear Creek cemetery. For the next sixteen years, until his death in 1984, he would take visitors out to admire his art. After working a lifetime in limestone, Elgar carved his monument in white marble and erected it on a base of gray granite.

"Dugan knew which stone holds up longest in the rain, and he wanted to be remembered for his carving a long time," said Morris. The air hammer in his fist, idle a moment, hissed quietly. "We all do,

Three Carvers

I guess. But the heck of it is, we don't sign our names to anything. There's people all over this country who walk every day past things I've carved, and they never heard of me."

During his four apprentice years, which he spent in this same mill, Morris cut stone for the Baltimore Cathedral. Later he worked on the National Cathedral, on churches and museums and hotels, on city halls, courthouses, gazebos, zoos. "I reckon my stone is spread over thirty or forty states."

A planerman who had been crouching behind a bale of sisal suddenly gave a yell and heaved a dust-covered sponge at Morris. Zipping through the air, it looked like a chunk of stone. Morris dodged, grinned, picked the sponge up from the floor and threw it back.

"These guys," said Morris.

"Do you keep track of all the jobs?" I asked.

"No, there's just too many. Oh, some are special, like the cathedrals. And some I keep pictures of—statues, fountains, the real fancy work." He pointed into the shadows at a concrete nymph. Stone dust in her hair and the folds of her gown made her look like an underclad girl stepping in from a blizzard. "Now you take that statue. The people at Busch Gardens bought it at one of these landscape stores, and they had me to carve them another one just like it in limestone. That was a good piece of work. You can go see it, in the Busch Gardens in Virginia." After finishing the Iowa Statehouse job, he would begin carving pieces for the restoration of the West Front of the Capitol Building in Washington. "I'll remember working on my nation's capitol, you can bet on that."

But he would spend half a year on the Corinthian caps for Iowa. Each would take him about ten days. By mid-afternoon today, despite his frequent breaks for cold coffee or tobacco or talk, acanthus leaves were already beginning to emerge from the blank stone. Once he had roughed out one side of this first cap, he would transfer the pattern to the blank sides by using a gawky contraption of jointed rods called a pointing machine. It looked like a robot's idea of a spider. Clamped in place on the carved face, it would enable Morris to map the exact location of high points and low points, and then to reproduce these contours on all the remaining faces.

Deftly, like a surgeon slicing flesh and with a surgeon's lordly bearing, Morris peeled the stone back, layer after layer, working down toward the image he carried in his mind. I watched intently.

"You want to try it?" he said, offering me the hissing hammer.

He showed me the proper grip: chisel bit in left hand, its butt end thrust into the socket of the pneumatic hammer; right hand wrapped around the hammer's shaft, thumb over the air hole to adjust the force of the pounding. The hose snaked along my arm and across the floor to a compressor. In my hand the tool felt like a real snake, twitchy and muscular, slippery, willful. On a piece of scrap I tried carving a straight line, but the hammer bucked in my fist. I tried it again and again, pecking out ragged curves and curlicues, over and over, exhilarated, stopping only when I knew I had kept Morris long enough from his work. The marks I made on the stone resembled the swervy tracks left by nightcrawlers on the sidewalk.

"It ain't something you pick up the first go," Morris said.

"Do you have any apprentices?" I asked.

"I've had them. But right now there ain't work enough to keep more than one man busy. Sometimes, if I get behind in a job, I'll call up one of the retired carvers and he'll help me out. None of the old guys ever really quits till he dies. But as for young bucks coming along, I don't see any. There's lots that would *like* to. But you can't make a carver without a lot of good jobs to cut, and that's just what we don't have. The way it looks right now, I'm the last there is."

Three Carvers

As quitting time drew near, the tempo of the mill slowed down. It was Friday, and the foreman came around to deal out paychecks. The man running the traveling crane—his cab draped with a Confederate flag, sun dashing from a gold ring in his left ear—lowered a can on a string to collect his pay. Hookers dropped their chains into buckets of oil to soak over the weekend. Sawyers hosed down their machines and swabbed the gears with grease. Morris gathered his tools into the battered box and snapped the padlock. With the air hose he blew dust from his clothes. Caught in a slant of sunlight, haloed in silver haze, he looked like a spirit visitor.

"Don't you boys wear out your eyeballs," Morris said. Lunch-bucket swinging, he headed for the parking lot and that luxurious new pickup.

With the machinery shut down and the men gone, the hush that filled the vaulted cathedral of a building seemed thick and creamy. After a minute of this smothering silence I began to hear the trickle of water through the floor drains, the skitter of cats sniffing around for tidbits the men had left, the cozy twitter of birds high in the rafters.

After quitting time at Woolery mill, on most of the days we visited there, one pneumatic chisel kept buzzing. The hand guiding it belonged to David Rodgers, a sculptor with a fierce enthusiasm for his art and for limestone. He might have arrived at mid-morning, and might not leave until dusk.

His wayward schedule was not the only item about David Rodgers that struck the Woolery men as odd. Off by himself in a corner, he never joked, never took breaks, never shot the breeze when you walked by. Sometimes he wore gloves, and always he protected his ears with muffling pads, like those worn by baggage handlers at airports. Not only had he finished college, he had finished graduate school, with a degree in fine arts. He didn't come from Indiana, not even from the Midwest, but from away out west somewhere. (Arizona, Oregon). Strangest of all were the things he carved. They looked like nothing on earth—like slices of the moon, maybe, curved and smooth and blank.

"I used to try talking with him," a shoveler told me, "but I never could figure out what he was thinking."

A cutter said, "I just can't *see* them things he makes. But people buy them to put out front of schools and banks and places, and you can't argue with that."

"What David Rodgers is," a mechanic declared, "is a forty-year-old hippy."

In scuffed leather boots, blue jeans, undershirt, his steel-wool hair cut short, gray beard trimmed neatly, gray eyes simmering and bright, he did not look at all like a hippy; but the label was a measure of how alien he seemed to the Woolery men. Unlike Jack Kendall and Henry Morris, Rodgers had never studied under a master carver, had never climbed his way up the mill's rugged ladder. He had arrived at a passion for limestone by way of academic training in sculpture. "I don't carve because my father did, or because I grew up admiring the old craftsmen. I carve limestone because it's a gorgeous material. I *chose* it." The names that surfaced in his talk—the stars he navigated by—were Michelangelo, Rodin, Henry Moore, Brancusi, Hepworth, Arp. For shapes he looked neither to medieval architecture nor to the Indiana forest, but down into the depths of our flesh—bones, wombs, birth canals—and into the hidden structures of nature.

We observed him working on an enormous sundial for Chicago's Lincoln Park Zoo. Chisel in hand, intense, mind and muscles concentrated on every gesture, he circled his great curving slice of moon not like a fencer but like a bullfighter. When we broke into his spell—Jeff with the bulky camera, I with notepad and raised eyebrows—Rodgers had to blink a few times to bring us into focus. Then he launched into a monologue. He unrolled a blueprint to show us how the finished constellation of stones would look. With a pencil he scribbled out the mathematics for us—latitudes, longitudes, equinoxes, angles of shadow, declination—explaining his sense of the sculpture's posture on the turning earth.

The airy mathematics he had encountered in college kept him from realizing his childhood ambition of becoming an astronomer. "If I couldn't study nature as a scientist," he said, "I thought I could study it as an artist." At first he had worked at painting the human body. Frustrated by two-dimensional canvases, he then tried to grasp the body's three dimensions by sculpting figures in clay, plaster, bronze, and wood. All the while, studying here in limestone country, he ignored stone completely. Then a visiting professor who had moved here for the sake of the stone brought him a mesmerizing gift from the quarries. "Ever since that time, limestone has been my favorite medium. But I had lived here for years without knowing this beautiful material or this great industry even existed." He bought his first carving tools from Harold Elgar, rented a house from a former cutter, and discovered how much knowledge of stone was bound up in the men and the mills.

Eventually he set himself up in a dusty corner of Woolery, among the plaster casts of saints and cherubs. His practice was to contract with the mill to rough out his puzzling, primordial shapes on saws and planers and lathes, and then he would finish each piece by hand. "I like working around big machines, and around men who know how to use them. I like feeling that my art is the fruit of an industrial process. Without all these men, without their tremendous skill, I couldn't do what I do. I need 150 or 200 years' worth of experience on the part of other men in order to produce one of my pieces." With a sweep of his arms he gestured at the empty mill. "There is an amazing concentration of knowledge here, a resource as remarkable as the stone itself. I don't think it should be allowed to die out."

He spoke with a visionary's smoldering zeal about his schemes for renewing the limestone industry. The owners should leap into high-tech, modernize, design graphics programs that would show architects a million elegant structures rising and falling on computer screens. The university should build a research park constructed entirely of local stone, and hire engineers to imagine new ways of using it. The state of Indiana should declare the men who dig and shape limestone to be public treasures.

"Of course, I'm partly selfish in motive. My sculpture is an ornament, an embellishment, of an environment. And ideally that environment should be courtyards and stairways and buildings of limestone. I want the industry to survive and become vigorous again for the sake of my art—but also for its own sake, because the stone men have accomplished great things."

The fate of the stone men reminded him of the Indian tools and petroglyphs he had discovered in New Mexico. "Here were human creations, thousands of years old, and they still spoke to me. The carvings in the cliffs showed me the timeless forms of nature. The scrapers and gouges still fit my hand. Stone had preserved those ancient visions and purposes." Stroking the pale flank of his Lincoln Park sundial, he said, "I hope the children climbing over what I've made, looking at the shadows and the play of light, will understand things about our place in the scheme of things, understand even without being able to say quite what it is they've learned. The sort of intuitive knowledge the Indians possessed."

10 Truth on the Back Roads

Walking toward the barricaded entrance of Indian Hill Mill, a ramshackle place on Clear Creek southwest of Bloomington, we were confronted by a tom turkey. Feathers boastfully erect, wattles bright red, great forked feet stamping the dust, he looked formidable enough to carve up a farmer for Thanksgiving. Jeff, a city boy, wondered aloud if turkeys were really as harmless as people claimed. Being a country boy, I reassured him. Turkeys are chicken, I told him. Whereupon Mr. Tom threw back his head and let loose a bloodcurdling gobble, a sound straight out of the brain's nightmare basement, and I flinched a giant step backwards. Jeff stood his ground. As if this cry had alerted the kingdom of poultry, we were quickly surrounded by a throng of geese, ducks, guinea hens, and chickens, each variety gabbling its own birdy dialect.

This squawky armada accompanied us uphill, gathering recruits as we climbed toward a red-and-white pole barn. We were guessing that whoever owned these birds might be working there and might give us permission to snoop around the ruined mill. Two pickups sat out front of the barn, one of them with the snout end charred black and the windshield shattered as if from an explosion, the other looking as scraped and banged up as a tank after a tour of duty. Also out front were chicken runs, rabbit hutches, the carcasses of dismantled machines, snow tires, a burst hydrogen tank (also fire blackened), buckets, feeding dishes, deflated sacks, a folding chair for catching the breeze. From the open mouth of the barn, which was nearly filled by a Coachman trailer, came the English accent of a BBC reporter describing a riot in Sri Lanka.

The poultry announced our arrival in several languages. The man who emerged from the barn to see about the racket seemed no more likely, on this backroad, than the British radio broadcast. His beaming face looked as if it had been scrubbed clean of doubt by decades of meditation. The untied laces of his tennis shoes flopped in the dirt as he approached. He wore, aside from the sneakers, only white shorts and a stealthy grin. Glimpsed across a poker table, such a grin might lead you to despair. Matching rugs of gray hair, thick and short, covered his chest and head.

When I asked him if we could look around Indian Hill Mill, he scratched the mat on his chest and said, "You can dance in there, if you want to, so long as you remember that I never saw or heard of you. I don't know you're here. You understand?"

We understood. But I was too intrigued by this gatekeeper to leave immediately for the mill. On the radio a Mozart string quartet followed the BBC news. The turkey and chickens and ducks and geese encircled us like an expectant audience at a theater-in-the-round. We gave our names; he kept his. The look he beamed at us made me think he knew a murderous secret—a bomb was planted in our car, a snake was breathing at our ankles—but had no intention of telling.

"Did you ever work in the mill?" I said.

"Nope."

"You farm?"

"Nope."

"That's quite a turkey," said Jeff.

"Yep."

The bristly bird, inflated by a sense of his own dignity, swelled even more. His wattles turned from crimson to blue.

"What's that change in color mean?" said Jeff.

The man rubbed a palm across the gray rug on his head. "It has to do with the circulation of blood. But what it means is anybody's guess."

"Do you sell poultry?" I said.

"I just raise them, the good-for-nothings. Gid-out!" he yelled, brandishing a sneakered foot as if to kick. The birds strutted nervously, flapping and scrawking, then resumed their watch. Repenting, he lifted a branch next to the burnt-out pickup to show us a gaggle of newborn ducklings. "They're French. Grade-A fancy." He let the branch fall and turned his sly grin on us. "You want to know what I'm doing way out here beside a run-down stone mill in the middle of nowhere, don't you? Well, I'll tell you." His eyes, never fully open, now clammed almost shut. "I make mountain dew."

During our rambles in limestone country we had run into scalawags of various descriptions, but not, as yet, a moonshiner. "Mountain dew?" I said.

"Come have a look." He waved us toward the barn.

Only the turkey, boldest of the birds, followed us inside. Next to the Coachman trailer stood a refrigerator, a table, and a stove, all hospital white. "I sleep in the coach and eat out here. Pretty tidy arrangement, wouldn't you say?" Pipes and hoses, covered by loose boards, passed under the walkway. The turkey's great splayed feet jounced the planks as he tottered after us. Our guide halted at the stove, opened the oven door, and with a sweeping bow said, "Climb on in, Tom." The turkey jerked his head up, swung around, and beat a hasty retreat into the sunshine, rattling the boards as he went.

"He's no fool," said the bird man. Beyond the stove, he drew aside a curtain and revealed a clicking, oozing, percolating laboratory that might have been a leftover set from a late-night movie. Red lights blinked. Liquid bubbled through glass tubes. Tall cylinders of concrete and steel, stirred at the top by electric motors, spewed thunderheads of steam. "Reactors," he told us. A gauge on the one nearest me read 420 degrees Centigrade. The air above our heads was crisscrossed by pipes and wires, the walls were encrusted with switches and relays and valves, the benches were obscured by glassy-eyed monitors: it was a plumber's and electrician's nirvana.

"I built every piece of this myself. Twice, in fact, because my first lab burned down a year ago."

"Did something overheat?"

"Only tempers. Some boys who'd been renting the old Indian Hill office got mad when I discovered their marijuana growing behind the mill in my blackberry patch. I could just see the sheriff stumbling onto that and paying me a visit. So I cut it down. A few days after that I went to check the mailbox, and in a little bit the boys came peeling by me, throwing gravel, and when I got back up here they had fires set in three places." This explained the charred truck and the burst hydrogen tank out front. All he managed to save from that first lab were a few cylinders of butane, which he had rolled away from the flames while a volunteer fireman doused him with a hose. "That was a little hairy. I wouldn't want to do that for a living."

Truth on the Back Roads

After the fire he thought about giving up and going fishing. But eventually, to avoid boredom, he rebuilt.

"Do you want to see my product?" From a desk littered with calculators, timers, logbooks, and twitching meters, he picked up not a bottle, but a manila folder. Opening it, he showed us a curving graph. "Paper, that's what I sell. That's what I make out here in the middle of nowhere—inkmarks on paper." (As a writer, I recognized the improbability of that.) Not a moonshiner after all, he was an organic chemist. On a notepad he began drawing molecules—benzene, butane, a flock of polymers—and writing formulas, all by way of explaining that he tested catalysts for use in the manufacture of marine resins.

I peered at the inky cobwebs, trying to recall enough of college chemistry to make sense of them.

"It may not look like much to you," the chemist said, "but right there is a little scrap of what God had in mind when he made the universe."

"Could I keep it?" I said, reaching for the paper.

"Secret, I'm afraid." He quickly folded the sheet and stuffed it in his pocket.

At every stage in our tour through his lab he seemed, like his turkey, at once boastful and skittery. I discovered at least part of the reason when he told us about his background. Born and reared in Indiana, he had worked as a chemist all over the country. He held thirty-eight patents through various companies. When a firm he worked for in Texas was bought by a Japanese company, he was pensioned off and forced to sign an agreement that he would not pursue his research on catalysts for anybody else. "So I came home to set up my lab in Indiana, where I figured they wouldn't care much about

STONE COUNTRY

Texas contracts." Cautious, he was only selling his results abroad, to companies in Germany, Switzerland, and France.

"But why did you settle just here, beside a stone mill?" I asked.

The chemist reflected on that. "I suppose because there's still a rich feel of men and machines around this place, a hum of work in the air."

Inside the twilit cavern of Indian Hill Mill, the dirt floor still held the footprints of stone men, interlaced by the meandering tracks of possums and raccoons. Shut down only ten years earlier, the building looked as though it had been neglected for a long while before that. Great swatches of clapboard had been torn from the steel frame. Patches of sky showed through the roof. Bushes and nodding weeds and head-high trees grew in the puddles of sunlight. Honeysuckle and poison ivy fingered in through broken windows. A scooped-out stone held the bones of a small kill—snouty skull, hooked canines, curving spine. On the walls electrical boxes gaped like giant clams, their knife switches corroded open. Much of the machinery had been sold for scrap, but the gang saws, a few planers, and a diamond saw remained, tinged all the colors of sunset, like engines rusting in a drowned ship.

Blocks, half sliced, had been left in the gang saws when the men quit work on that last day. Wooden triangles wedged them in place, to make sure the cuts would be true. A pair of overshoes touched noses beside the sandpit. A flannel shirt draped over the arm of a sawyer's chair. Plastered across an engine housing, a sticker from a local gun shop advised: TAKE YOUR BOY HUNTING INSTEAD OF HUNTING YOUR BOY. (When that sticker was put up, many sons were fighting strangers in the jungles of Vietnam, and some were fighting policemen in the streets of America.)

Truth on the Back Roads

The feel of vanished men was strongest in the blacksmith shop: battered forge, drill press stripped back to the dumb chassis, broken drive belts curled on the floor like sleeping pythons, barrels full of bolts as thick as a baby's arm, buckets and wooden crates and sagging file drawers crammed with parts, gloves scattered on the floor like discarded hands. A newspaper of August 12, 1974, lay open across a stool: BENCH DRIVES IN FIVE AS REDS MAUL METS. The wooden doors of a tool cabinet, its handles made of welded chain, were twin palimpsests, scrawled over with calculations in red crayon and black pencil. Screwed to the inside of one door, tilted so as to reflect the mug of a mechanic, was a car's circular rear-view mirror.

Altogether I found six mirrors hidden away in the recesses of Indian Hill Mill, most of them jagged shards, held in place by wires or bent nails. I thought about the men at day's end blowing dust from their clothes with air hoses, smacking dust from their caps, thumbing dust from their eyes, stooping down to watch themselves in cracked mirrors raking wet combs through their hair.

Every inch of our looking made me ache with a sense of the dispersion of knowledge, skills, and lifeways. All the while, outside, the chemist's poultry creaked and croaked like rusty pumps.

The bird that greeted us out front of Bob Thrasher's house, five miles west of Harrodsburg on Popcorn Road, was a magisterial goose. Less than a fourth the size of Tom Turkey, it was twice as bold. Honking, wing-flapping, nipping at the red laces of my boots and the creamy laces of Jeff's running shoes, it herded us toward Bob Thrasher, who sat in a wheelchair in the shade of a tulip tree. He was a tight-muscled man of sixty with a flap of brown hair and a sunbaked outdoor face. The inward sag of his mouth and the squint in his blind right eye gave him a crooked smile, as if he were bitterly amused by the world's antics. His right leg, crushed three months earlier under the wheel of his tractor, was propped on a stool. That pant leg was cut off at mid-thigh and his foot was bare. As we sat down to share the tulip-tree shade with him, he was rubbing the swollen knee.

"You ought to see it in the X-rays," he said. "It looks about like a junkyard in there."

He numbered the breaks for us, from ankle to hip. After the tractor had passed over him, he had dragged himself to the edge of the field and lay panting, in and out of consciousness, for thirteen hours, until a neighbor found him. "I got no business being alive. I've done so many fool things in my life, I'm overdue. I tell you I did wonder, laying there, if this was my time come at last."

We had talked with him before the accident, about his work in the mills and about growing up in a stone family, but this was our first visit since his return from the hospital. Before, his talk had crackled with the fires of irreverence. How would it be today, when he was in constant pain, when he faced the prospect of never again being able to walk?

"All busted up like this I can't do anything but read books and talk. Lord, the things that need doing! I must have twenty miles of fencing, and half of it's crying for work. I've been having to read my cows a legal description of the property to keep them up. And you know how dumb cows are. They don't pay me any mind. And do you know why cows are dumb? Why haven't animals taken over the world? I'll tell you. What do you do when you get a smart cow, one that can open a gate, say, or get into your feed? Do you keep her? No sir. You butcher her, is what you do, and you breed the dumb ones. It's downward evolution."

The goose, at least, had a mind of its own. Having untied my boots, it was nibbling at the seam on my bluejeans, working up toward the knee, and no amount of shooing would deter it.

"My son finished getting in the corn," said Thrasher, "and his wife mowed my clover for me."

The corn, higher than our heads now in August, reared a wall of green between us and the road. The second growth of clover was already tall enough to hide a cat. In fact, it did hide a cat, a gray veteran that slouched toward us with a field mouse drooping from its chops like a handlebar moustache. The front yard was a highway for pets: in addition to the goose and cat, our visitors included three ducks, four chickens, and a galumphy black dog. Had the fences been as bad as Thrasher claimed, we would have been visited by saddle horses and beef cattle as well. Their pastures rolled away behind the house; beyond the pastures lay woods and fallow fields. Thrasher's five hundred acres stretched more than a mile in from Popcorn Road, stretched all the way to his father's place on the next road to the north, a road named Thrasher.

As far back as anybody could remember, the Thrashers had farmed in the vicinity of Harrodsburg. "My father was the first one who couldn't make a living at it. I can shut my eyes and see him going off to work in the mills when I was only a toddling boy. I must have been four or five years old, which would make it about 1928, 1929. Stone opened up all around here. Quarriers walked past our gate coming and going, lugging their lunch pails. When they stoked up the channelers first thing in the morning, the smoke was so thick you couldn't hardly see to walk down the road."

By the early thirties, work in the stone belt grew scarce. The unions lost their leverage. Men hungry for any sort of a job would sit around the lip of a quarry hole, waiting for somebody down there to make a slip. If a quarrier got hurt or mouthed off, the foreman would send him packing and crook his finger at one of the watchers. There was no insurance, no pension, no sick leave. If you could move, you worked. If you missed a day, no matter why, you lost your pay. Fortunes were made and mansions built out of the gristle and ache of men like Thrasher's father. Recalling those days, Thrasher simmered with a wry anger.

"There was a few families with money, the owners, and there was a whole lot of end-of-a-mud-road families like mine. The money folks didn't know the first thing about us, and we didn't care much about them. Most of them saw us as a bunch of ragtags, one the same as another, like a hammer you pick up and use awhile and then lay down."

Being ragtag and poor, shy on book knowledge and little traveled, the stone men were not very tolerant of outsiders, a category that included people with dark skin and foreign accents. Thrasher could remember hearing, as a boy, about the fiery threats and ghostly assemblies of the Ku Klux Klan, and about a vigilante gang called the white-cappers, who dragged men they did not approve of into the woods and beat them by way of instruction. (Older stone men had told us about seeing crosses flame in the town parks of Bloomington and Bedford, and on the lawns of families who broke the unwritten code.) Italians and Germans, often imported as strike-breakers, were given a hard time in the mills. Blacks were given a hard time everywhere.

Thrasher also remembered listening, as a boy, when his father stopped at the gate after a day in the mill and shot the breeze with the other stone men. "But Dad never wasted his joking on me. With me it was always work, work, work. From about October until spring, every weekend I lived on the end of a crosscut saw. I hated that. I flat *hated* that. But my dad farmed and worked in the mills all summer, so we had to get in our wood in the winter. Since he didn't quit until after dark, I had the job of going to hunt the cows. We didn't feed them nothing when we milked them. Just milked them, you know, and turned them out. So naturally they'd go hide. We had bells on them, but they'd stand just as still, like a field full of statues. They'd go as far from the barn as they could get, and I'd hunt for them in the dark, barefoot, stickers and thorns everywhere. Finally I'd hear a tinkle, and find them. I trained a red polled cow to let me ride her back to the barn."

Thrasher himself started farming at sixteen, went to fight Hitler at eighteen, came home, numbed and searching, at twenty-two. "I was a scout in Europe, which is like being a duck in a shooting gallery. My buddies got killed right and left. For a long time after I got back, everybody I knew my age was dead. It was like I was stealing from them, just being alive. I felt awful. I'm not sure I ever did get over that."

Married, with children coming, he worked for a while at the General Motors foundry in Bedford, but never liked it, despite the good pay, because it was too regimented, too rush-rush, too much of a head-bobbing and order-barking place—too much like the army. As soon as a job turned up in the mills, he took it, and stayed around stone for the next ten years, running a diamond saw and a split-face machine, hooking for the crane, driving a truck and a forklift, anything and everything. "There was an ease to it that I liked. If there was heavy work to do, you worked hard; if it slacked off, you told stories or shot craps or put up your feet and read." Running a diamond saw in the evenings, he plowed through biographies, histories, novels, and all sixty-six books of the Bible while the circular blade plowed through stone.

"It wasn't all sweetness and light. We were an ornery bunch. We'd put loose bolts on the railroad tracks and stand back to watch the boys on the flatbed cars go thump-thump-thump, jouncing over them. We had this foreman, Henry Brunner. He was a religious man, wouldn't cuss no matter what. So we'd do everything we could think of to put old Henry's religion under a strain. Like we had this one-hole crapper in the woods. Stank so bad I wouldn't go in it unless I was desperate. But there was this one guy who'd just *live* in there, instead of working. Well, one day we set it afire with him in it, to smoke him out. There was so much wet paper in the pit, the crapper burnt for three days. Old Henry, he just hopped in a big circle around the fire, yelling, 'Mercy, mercy!'

"And there was this guy, Clyde, drove one of the big tractors. We didn't like him. Had his nose in the air. One night we took and unbolted the plate from under the motor on his tractor, put a dead cat inside, then bolted the plate back on. Next day, as soon as Clyde got his engine all hot, it started to stinking. There he sat, sniffing, and couldn't figure out where the stink was coming from. Henry Brunner hollered, 'My goodness, it's terrible, it's horrible, what on earth are we going to do?'

"Another time we had this boy, half crazy, he'd park his car in the same spot every day, under a big tree. Soon as the whistle blew at 3:30 he'd run jump in his car and gun the motor and peel gravel leaving the place. So one lunch time we took and tied a fifty-foot length of cable round that tree and onto the frame of his car. When the whistle blew and he jumped in we all looked kind of sideways at him. He went peeling off, as usual, for fifty feet, and then he stopped all sudden and nearly banged through the windshield. Henry Brunner came hopping and yelled, 'Sakes alive! You 'bout to murdered the boy!' That was the sort of pranks we used to think was funny."

Except for a knife of pain in his knee, which clamped Thrasher's mouth shut every now and again, only the need for air interrupted his stories. Jeff had his photographer's eye cocked at him, posing the gimpy farmer against the stand of corn. I listened with pleasure to the turns of his sentences. The goose snipped at the brads on my hip pockets. Dog and cat and ducks and the rest of the menagerie waddled or loped or clucked through our triangle of chairs.

"The owners didn't waste any money on us," Thrasher said after the knee eased up on him. "Like on heat, for instance. In the winter, you'd turn on the radio first thing in the morning to see what the temperature was. If it was above eighteen degrees, you'd go to work; if it was below, you'd stay home. Not for our sake, you understand, but for the sake of the stone. The thing was, if it fell colder than eighteen degrees, the pieces would stick together.

"But the owners weren't all skinflints and stuffed shirts. Some of them would treat you like a human being, almost like an equal. I remember I worked in a mill that belonged to this old coot, Daddy Beck. I was outside one beautiful morning in the fall, loading split-face, and Daddy Beck drove by in his big old Buick and called out to me, 'A day like this makes you glad to be alive, don't it?' And I knew just what he was feeling. That was about as good as it ever got between us and the owners."

Pain was no novelty for Thrasher. His right eye, puckered nearly shut, had been blinded at the mill when a piece of strapping broke and raked across the cornea. The nails on his propped-up foot were yellow and thick from having been mashed hundreds of times by dropped tools and rocks. The skin of his hands was embroidered with scars.

"What I liked about the mills was the independent sort of men who worked there. All of them had something else they could do—farming, carpentry, plumbing, fixing cars—so they could afford to be cantankerous. They didn't turn sour when they got laid off. We'd all go on strike pretty regular. If it looked like it'd be a long one, I'd just plant more corn."

By 1961, when Thrasher had four children in college and a hefty mortgage, the lay-offs and low pay—"I'd work every day I could, maybe twenty-five weeks a year, and earn $3000"—forced him to leave the mills and get a cushier job. He moved to the post office. There he stayed until his accident. Nearly a quarter century of working in an office, most recently as assistant postmaster, had not increased his regard for paper shuffling. "The way it is with most jobs today, it don't matter a whole lot if people even show up for work. If ninety percent of the people was to stay home tomorrow, it wouldn't make much difference. Where the world's headed, if you ask me, is everybody standing around a Xerox machine with a cup of coffee in their hands, waiting their turn in line, and somewhere there's an old guy raising a zucchini or something to keep them fed."

Clasping both hands under his junkyard knee, he shifted his leg on the stool. You could estimate the number of breaks by counting the grimace lines on his face.

"That's why I always come back to farming," he went on. "With farming, you know when you do something it really matters. If you don't do it, nobody else is going to do it for you." He stared at the knee, silent a moment, then added: "Except maybe one of your kids will help you out when you're banged up, or your wife, or your daughters-in-law, or your dad." His father, eighty-five, was still farming. Running a dairy herd for thirty years, he'd only missed one milking. "He could still paint a flagpole if he wanted to."

I wondered if Thrasher himself would ever climb out of that wheelchair to drive a tractor or mend a fence. The land sloping away from us in all directions bore the marks of his labor, and he bore the marks of the land. He had built the house of frame and brick, the barns, the tractor shed, had planted the tulip tree we were sitting under, had shaped the woodlots, seeded the pastures, nursed the topsoil. He'd chosen this parcel of land to farm because his father, back in the dirt-poor days, had rented it from a well-to-do man in town. "When I was a boy my dream was to buy this place, because my dad worked on it so hard without owning it."

The farm straddled two townships, Clear Creek and Indian Creek. Thrasher's wife was trustee for one of them, his son for the other. "Around here that means something. Township trustee is a harder-fought election than president or senator. It's neighbors voting on neighbors." The same neighbors had elected Thrasher to the county school board over and over, for eight years. They'd called him a Communist when he joined one of the first delegations to visit China after the Nixon thaw. Their puzzlement about him had only deepened when several emissaries from China made their way out Popcorn Road to have a look at his farm. (One of these visitors, a journalist, wrote a story about Comrade Bob that was read by several hundred million Chinese.) Baffled by this man who read thick books and cultivated corn, who milled stone and traveled around the world, who spoke sometimes like an end-of-a-mud-road hick and sometimes like a white-collar gent from town, the neighbors called him an intellectual peasant.

It was time to go. I could hear the pots of dinner clanging in the kitchen. The goose was feasting on my ankles. Thrasher's face sagged from pain and memory. I ruffled shut the leaves of my notepad. He glanced at it, yellow sheets smeared with ink, and said, "I oughtn't to be telling you all this, but it's only the truth."

11 Stone Towns and the Country Between

Water-carved chunks of limestone flank driveways and rise from flowerbeds, pale and knuckled like the vertebrae of dinosaurs. Landscape architects send trucks long distances to fetch them from quarries and creekbeds, but the local people drag them home behind tractors or in sagging pickups. The backroad name for them is "waterwarts." The prize ones are pierced with holes, letting air and sunshine pour through and making the massy rock seem riddled with light.

In the tulip bed at the center of Jim Medley's driveway turnaround, out Red Hill Road from Ellettsville, a fan-shaped waterwart stands on edge. The hole through it, large enough to swallow a cabbage, gives it the look of a great staring cyclops. "Daddy calls it his potty stone," Medley tells me. "If we still had a privy, I believe I'd try it out. You'd never find a smoother seat."

Shaped like a dolmen, the sign welcoming you to Ellettsville is a rough-hewn limestone slab balanced across two stubby uprights.

Stone Towns and the Country Between

Carved beneath the town name are the words: "Builders of American History."

A hand-lettered notice in the town's main pizza joint proclaims: NO TOBACCO CHEWING. A wooden peacock with a spray of real iridescent feathers hangs on the wall alongside posters of rabbits, clippings about local football games, and baskets filled with dried flowers. The cigarette machine supports a terrarium and a clutch of baseball trophies. At lunchtime the men play Donkey Kong and Pac-Man on the video boxes while their orders cook. They eat from varnished picnic tables, and keep their heads warm under visored caps: FARM BUREAU CO-OP, TRUCKERS OF AMERICA, JOHN DEERE, ODON CLOTHING COMPANY, LUBRIPLATE, HYDROPOWER, MAIL POUCH. In the parking lot, the bumper stickers on their trucks say: EAT MORE POSSUM, KIDS NEED TO BREATHE, I'D RATHER BE DANCING, IF GUNS ARE OUTLAWED ONLY OUTLAWS WILL HAVE GUNS, NUCLEAR WAR IS BAD FOR CHILDREN, NUKE THE WHALES, I LOVE GRANDMA, HAVE YOU HUGGED YOUR DAD TODAY?

Where Jack's Defeat Creek crosses Ellettsville's main drag, the Village Inn greets you with a round clock labeled "Time to Eat." Indoors, a sign over the cook's window gives you a cheery "Hi, Neighbor." The walls are decorated with the horns of bulls, wickerwork fans, and placards bearing words of wisdom. One saying comes from Will Rogers: "Politicians are the best men money can buy." The waitresses are grandmotherly in age and manner, making you feel like a long-lost grandchild who's dropped by for a meal. Their hair is piled in billowy mounds the color of gunsmoke. Their hands and voices are in no hurry as they serve you "The Stone Cutter's Breakfast"—omelet, butt steak, fries. For dessert they bring coffee thick enough to hold the spoon upright, and a wedge of peanut butter pie.

STONE COUNTRY

On the highways, billboards promote fast foods, herbicides, pesticides, fertilizers, cigarettes, Army, Navy, National Guard, Savings Bonds, booze.

Signs along the edges of cornfields announce the brands of seed: PIONEER, BIG-D, TROJAN, AGRIGOLD.

Religion leaps at you from the roadsides: FAMILIES THAT PRAY TOGETHER, STAY TOGETHER; JESUS SAVES; GO TO CHURCH THIS SUNDAY; GET RIGHT WITH GOD. Words painted in day-glo yellow on the window of a gas station holler two hard questions:

    IS YOUR   IS GOD
    BATTERY   IN YOUR
    SAFE?   LIFE?

A sheet of plywood lettered in red and nailed to a tree demands:

    HURTING?

then answers its own question:

    GOD CARES.

The churches come in more brands than hybrid corn: Primitive Methodist, United Methodist, Regular Baptist, Separate Baptist, Southern Baptist, Old Dutch, Seventh-Day Adventist, Whole Gospel, God of Prophecy, on and on. A street sign on the drive leading up to a pentecostal assembly in Peerless advises:

    DEAD END
    CHURCH.

Two photographs have been thumbtacked to the bulletin board just inside the door of Summitt's Grocery and Post Office, the only store in Stinesville. One shows a candidate for sheriff, grinning around a fat cigar; the other shows the local raccoon-hunting champ, his feet hidden by an avalanche of ring-tailed pelts. Beside the photographs hangs a calendar with the names of villagers printed on their birthdays and anniversaries.

Whatever anybody in Stinesville has to say eventually gets said at the rear of the store on a bench between the woodstove and the cash register. I sit there, jugs for ears, beside the proprietor. Forty-nine, with a steel-colored brush of hair like a scouring pad, Bob Summitt possesses the rounded bulk of a man who has sat for quite a few years within easy reach of ice-cream bars and soda pop. Today he is red faced from having sat all morning on a mower. As we talk, he stokes himself periodically with pinches of tobacco. The soft-drink cooler is at my elbow, and every time the door opens I catch a refrigerated breeze.

It opens often this July afternoon. While Summitt rests up from mowing, his wife waits on the needs of the town. Jumper cables, cheese, a stump puller, bread, pint of chain-saw oil, fire extinguisher, aspirin, sliced ham. She is a brisk, soft-spoken woman, her steel-gray hair cut short to match her husband's, a pair of black, round-framed glasses giving her an owlish look. Whenever box-holders come in, she fetches their mail without being asked. People sort out the letters from the bills on her counter and sit down beside us to read anything noteworthy. They also read the newspapers that are stacked on the bench, or catch up on the news in the perennial fashion, by talking.

Summitt began work in the mills as a laborer, shoveling stone dust and scraps. "No matter who you was, you started out with a shovel in your hands." Eventually he apprenticed as a planerman, but one week after completing the apprenticeship he was drafted. By the time he came home, in 1959, his hands had lost the feel for tools, and business in limestone had sunk very low. The old men, with decades of seniority, would not give up their places. Young men like Summitt, in order to work at all, had to shuffle from mill to mill, and had to put up with the shabbiest equipment. "When I quit working stone for good I was down in Indian Hill, and they put me to using these wore-out steel tools. Three passes, and they needed sharpening. I'd never worked with anything but carbide tools, and down there only the old guys had them. If they wasn't using them, they'd hide them in the dirt just to keep me from getting my hands on them. Well, one day I was using them soft tools and broke three pieces of stone, and I got so blame mad I never went back." It's no wonder that stone has gone under, he says. The owners never modernized. They stuck with the old equipment and old methods while other industries were keeping up with the times.

He knows for a fact that stone from his planer was used in a lot of famous buildings, but can't name a single one to save himself. The names were all there on the job tickets, but he hardly ever bothered to look. "To tell you the truth, what I looked for was 3:30 in the afternoon and a paycheck on Friday." After quitting the mill he worked all over, doing everything under the sun. Eighteen years ago he took over the store and post office. "And if I knew who my boss was I'd put in for retirement."

In two hours of sitting there, I meet two dozen people, every one of the men a former stone worker, every woman the wife or daughter of one, every kid the grandchild of one.

"When I was a boy," says a butter-voiced man with a cowpoke's face, "the quarries and mills were all there was. That's where your daddy worked, and that's where you worked, as soon as you got big enough to work at all." He lied about his age so he could start young, at sixteen, hooking stone. "Now there's a job where you've got to trust people. So long as the guy at the other end of the block has dug his hole and set his dog right, and so long as the derrick runner and powerman do what they're supposed to do, you're just fine. But if anybody blinks, you could be dead."

Paydays were celebrated with poker and craps, the cards played on stone, the dice on bare dirt. Many a week's wage was lost in a single roll or deal, and if the loser was married he would catch sand and hellfire when he slunk home. Once, a man who had grown too old for the quarries came back for the weekly game. After a year's work at squaring logs for railroad ties, he had saved up twenty dollars. In three rolls he lost every nickel. Whereupon he stood up, dusted off his hands, and declared, "Easy come, easy go." The police raided the game one time and made a big stink about it. The next week a new spot was cleared in the woods and the game resumed.

After the quarries near Stinesville closed down in the early 1960s, the gambling continued, but the paychecks came from elsewhere. Leaving stone, the men went to work fixing cars, airconditioners, televisions. They patched holes on the highways, pumped gas, screwed parts together on assembly lines, swept floors, drove trucks, whatever they could find to do.

When a mousy old gent creeps in to buy an ice-cream sandwich, he mumbles that he doesn't know anything about the quarries. He's well up in his sixties, toothless, wearing a moldering Chevrolet cap and the mournful expression of a beagle. His eyes aim at the floor, as if on the lookout for dropped keys.

Summitt prods him gently. "Who you think you're fooling, pretending you never been in a quarry?

"Oh, sure," the old man concedes, his voice a faint trickle, "I worked for a few years on a channeler." Like most of the stone men, he's hard of hearing. He whispers because he is afraid of shouting the way deaf men do. I lean close to listen. As a boy, he used to carry his father's dinner bucket from the farm house to Wallace Quarry. There was a regular beaten path that all the children followed. His parents knew exactly how long it should take him to cover the distance, and if he was late at either end he caught the mischief. Since the trail cut right through a strawberry patch, in the season when the berries came ripe he ran most of the way in both directions, to leave himself time for picking and eating.

Along with three of his four brothers, he fought in Hitler's war. One of them was killed, shot to pieces in Italy. When the casket was sent home for burial, the surviving brothers knew there was no body inside. "I saw them battlefields. They buried whatever scraps they could find right on the spot." But the boys assured their mother that the slain brother was in the box, safe and sound. Almost crazy from grief, she wanted to open the casket to make sure. "That's why they send an MP along, to keep people from prying it open and finding out the truth."

Every man I've talked with today spent some time in uniform, usually the Army's olive drab. Some of the old mossbacks fought in the Great War; the balding men fought in the Second World War and Korea; and the young bucks (whom we can see buzzing their motorcycles and jacked-up cars past the window) fought in Vietnam. "You didn't think about whether you was going to go," the butter-voiced cowpoke tells me. "It was like starting in the quarries and mills at sixteen or eighteen. When you come of age, you just signed up and they handed you a gun."

Across Main Street from the store, in a tiny park, there is a war memorial, with enough names on it for a town five times as big as Stinesville. From over here you can see that Summitt's Grocery and Post Office occupies what used to be the Odd Fellows Oolitic Lodge. Built of wood in 1891, and after a fire rebuilt of limestone in 1894, it is linked to three other abandoned stores, all dating from the 1890s, all bearing the names of various Easton brothers. For obscure reasons, the Eastons opened four dry-goods stores side by side. During the town's heyday, between about 1890 and 1916, some twenty businesses flourished here, and the population crept over seven hundred. There were four taverns, and trains brought thirsty folks from all the surrounding—and tavernless—villages. The streets were so thick with people you could hardly walk. But in 1916 the town's largest mill burned down, and Stinesville's fortunes began to slide.

Many of the citizens eventually slid into Mt. Carmel cemetery, west of town. Except for a single wedge of pink granite—which looks out of place, like a cardinal in a flock of sparrows—every marker is limestone. Most were professionally carved, but a few, for children, were hacked out by hand. There is the usual forest of sculpted tree trunks with the limbs lopped off. On one of these a mallet and sledge have been carved right onto the branches, as if forgotten there when the job was done. The marker for the Titzel family is capped by a steam locomotive, complete with coal tender, all in lichen-covered limestone. In the sinkholes of the cemetery grow towering arborvitae, spindly cedars, blackberries, frilly topped wild asparagus. The weeds are knee high, except right near the stones, where lily of the valley and periwinkle hold their own. The weeds keep me from seeing a fat blacksnake until I've almost stepped on it. I freeze, and his lordship slithers under a memorial slab.

Many houses in the stone towns have an evolutionary air to them, no longer as fine as they once were, not yet as fine as their owners hope to make them. Warped clapboard shows around the edges of aluminum siding. Porch roofs lean on two-by-fours, waiting for fresh mortar on their limestone posts. Replacement windows, smaller than the

STONE COUNTRY

originals, wear collars of milky plastic. In winter, bales of hay snug up against the foundations for insulation; in summer, the hay mulches the garden. Silvery new stovepipes rise next to crumbling chimneys. The woodpiles are half the size of the houses. Gnarled antennas dangle from roofs, while in front yards the gauzy dishes of satellite receivers snag images and sound from all the world.

Yards in limestone country are decorated with plaster donkeys pulling carts, mirrored balls in tints of green and blue, wagon wheels, plastic flamingos, figurines of black servants (some with faces painted white) holding lanterns, statues of alert deer with ears pricked forward and tails uplifted, wooden ducks whose wings revolve in the breeze, model windmills with spinning vanes.

Gaunt rusting towers of real windmills loom between farmhouses and barns. The houses are frame, blazing white in the sunshine, rising two stories above limestone basements to gables filigreed with arabesques of wood. Metal-roofed porches clasp the ground floor in shadows. The barns are hump-backed, tin-roofed, with sides the color of burgundy. The cornbins are tall gleaming cylinders or squat blue tanks, their pointed roofs giving them the look of rockets on launch pads. Round bales of hay clump together in the fields, as if for company. In the farm lots women hang up clothes and men thrust their arms into the entrails of machines—tractors, bulldozers, trucks, cars—and kids make up games out of sticks and stones and dirt. Middle-aged cars, cannibalized for parts, rest on blocks of stone in the side yards. Elderly cars, gnawed down to fenders and chassis, sink beneath briars in the gullies of pastures.

The old state road heading north from Oolitic has been cut in two by a quarry. The townspeople must detour miles around the hole to visit their neighbors in Needmore. But they are used to shaping their lives to suit the needs of the great god limestone. The name of their town derives from the egg-shaped particles—the oolites—they saw in the rock. At one time nearly every house here belonged to a family whose menfolk worked in the quarries and mills. Too poor to build their houses out of anything except wood or tarpaper or concrete blocks, they used limestone only for stair treads, sidewalk flagging, pillars on porches, lintels, benches, birdbaths.

Since the road was cut, most of the gas stations have given up the ghost. Hoosier Avenue Liquors has taken over one of them, and appears to be thriving. Another one houses Stone Capitol Pest Control. And still another one, across the highway from the Living Word Church, is occupied by Stone Kingdom Realtors. The sign out front shows a pyramid with a mystical eye at the apex, but the realty business has winked shut. From Oolitic Rent-All a radio blares, advertising *Hot Rod Magazine*'s Super Nationals auto show in Indy, a "picnic of power" designed to "celebrate America's Love Affair with the Car." A few steps away from the Living Word Church, kids scramble and yell in the school yard. The boxy school itself hunkers down in a smear of blacktop, looking grim and bleak, like a grout pile at a quarry. Bas relief panels up near the roof depict a globe, an open book, a lamp of learning—and a basketball passing through a net.

Many of the women of Oolitic, having learned from the Bible that their hair is their glory, walk down the sidewalks under mountainous beehive hairdos. They look as though thunderclouds have gathered on their heads. In a town with one grocery, they can choose from four or five beauty parlors, including Raintree Hair Company and Tigress Lair Salon.

Banners draped from lightpole to lightpole across Main Street advertise the Oolitic Summer Festival. The newest attraction stands out front of the town hall: a pasty-white limestone statue of the World War

STONE COUNTRY

II comic book hero, Joe Palooka. About eight feet tall, dressed in boxing gear—shorts, cape, high-laced boots, taped fists—he's a big-chested brute, rock ribbed and square jawed, with a shock of hair draped over his forehead and bullet holes for eyes. For years he stood south of Oolitic in a park run by the Fraternal Order of Police. But the cops couldn't keep vandals from defacing him, so he was moved into town for safekeeping. In preparation for the summer festival, he's been sandblasted and his muscles have been touched up, but he's still an ungainly brute, long trunked and short legged, a Neanderthal in limestone.

Zigzag country roads follow the section lines, turning ninety degrees at the corners of fields. Roads topped with crushed limestone turn blazing white in summer, like pathways of snow. Driving them, you trail plumes of dust behind. In winter, under real snow, the roads merge with the fields. You keep on the hidden gravel by aiming midway between the fences—barbed wire strung on creosoted poles, rusty wire stapled to fat stumps, single strands charged with electricity, mossy limestone walls laid up without mortar, split rails, hedges. Cedars raise their green flames above the fence rows and flicker in the brown woods. Along frozen creeks the sycamores bristle like pale whiskers in black beards. Last year's stubble shows in dotted lines across the corn fields. Crows pump by overhead on raucous errands, the pitch darkness of their feathers deepened by the white of winter. Flocks of starlings dash ink strokes against the sky. Farmhouses float in the oceans of snow like islands, like mirages.

In front of trailers, low-slung ranches, prefabricated boxes, and suburban villas—homes whose owners punch the clocks in town—memories of farming still show in the choice of mailboxes: some are shaped like miniature barns, some like silos, some like corncribs; they're propped up by milkcans, hand pumps, plowshares, augers, upended cultivating disks, horseshoes welded into iron lace, chain welded into serpentine curves. Inevitably, some mailboxes also rest on pillars and piles and pyramids of limestone.

Some roadcuts near Red Hog Hill, south of Bloomington, were dug not with drills and dynamite, but with channelers, because the limestone was too valuable for blasting. In roadcuts near Harrodsburg, rock hounds with hammers and battery-powered drills prospect for geodes. Where the geodes are thickest, they have burrowed in several feet. Every now and again the overhanging ledge collapses, strewing the highway with jagged debris. Recently a chunk the size of a dump truck tumbled down and blocked two lanes of traffic. The rock face is sixty or seventy feet high, eroded at the top. Mud seams creep down between the upthrust knobs of stone, pouring out rust-colored deltas of clay. Crown vetch, pink with flowerheads in June, and yellow clover snag their roots in every smidgen of dirt. High up in the roadcuts the crystalline faces of exposed geodes shine from the gray rock like diamonds set in tarnished silver.

On the highway leading into Bedford from the south, the Chamber of Commerce has erected an antique derrick. The billboard affixed to the mast announces that you are about to enter "The Limestone Capital of the World." Back in the boom years, there were thirty or thirty-five mills cutting stone in Bedford; today there are two or three. Despite this dwindling, the townspeople still call the place Stone City, and the high-school teams still go by the nickname of the Cutters.

In the middle of June a carnival forms a ragged noose around the courthouse. Hawkers invite you to try your hand shooting basketballs and airguns, have your fortune read, drive leashed motorcycles or dented dodge-um cars, fill your stomach with sugary goo and then empty it again by riding the whirligig rides, all as part of the Bedford Limestone Festival. But there is no whiff of limestone anywhere in the festival itself, only bored barkers, horseshoe contests, a beauty pageant, a celebrity bake-sale, a bike-run-trot-walk fundraiser for a local child, and, to finish it off, a buffet dinner and dance at the Elks Club.

On the courthouse square the banks look prosperous, but other enterprises are ailing. Across the intersection from the Stone City Bank is the Greystone Hotel, once a hive of limestone wheelers and dealers, once a place where moneymen made decisions about the skylines of cities, now a dreary hulk with a caved-in roof and a lobby—glimpsed around the NO TRESPASSING signs—buried in fallen plaster. Nearby is the ruin of a fire-gutted cinema. You walk under the marquee, through a gap where the popcorn machine used to be, under a web of twisted girders and into open air. The emblem hanging in the doorway of the pillared Masonic Temple is broken and dangles askew. Three plaster owls, dyed an implausible brown and blue, have been stuck on window ledges up under the portico to discourage pigeons. But the pigeons flock around and bespatter the temple anyway, cheeky birds, oblivious. Around the square a few shops get by: The Ice Cream Klinic, Kloset Boutique, Smoke Stick Gun Shop, The Corral Video Games ("No Foul Language"), Greystone Gift Shop, Living Waters Ministry, Puffy's Tavern, and a bakery just called—in red neon—Bakery. But many stores are empty, their windows covered with paper or glazed with paint, and several display flame-red GOING OUT OF BUSINESS signs.

Railroad tracks run along the west side of the square. When a train lumbers through, the town stops dead. At one time, hundreds of flatcars loaded with stone rumbled over these tracks every week. Now the trains haul aluminum, plastics, automobile parts, livestock, coal. The milled stone that still leaves Bedford rides by truck, because the railroad beds are so broken down they will not bear the weight.

The county museum, in the basement of the courthouse, devotes a single glass case to limestone (there are a dozen cases devoted to dolls and cutlery and high-button shoes; there's an entire room stuffed with the memorabilia of war), and even this display is all but hidden by a photocopying machine, which the genealogists keep busy churning out family trees. In the gloom you can see a handful of limestone artifacts: a wreath, an angel, an eagle, a bust of Will Rogers, a plaque showing an Indian in a cactus landscape, and several of the pointy balls—like the bristling seedpods of sweetgum or sycamore—that apprentices carved to prove their skill. The finest piece in the case is a miniature statue of a stone cutter, holding a mallet in one hand and a chisel in the other. The sleeves of his jacket are rolled up over ropy forearms. He wears a railroadman's cap, an apron, a handlebar moustache, and the jaunty look of a man who knows exactly how far he is from the center of the universe.

Walk south from the courthouse square to Green Hill Cemetery and you will find a full-scale version of that miniature. Lifted high on a pedestal, the figure of a cutter lords it over the acres of graves. Not a memorial to anyone dead, the statue is a tribute to all diggers and cutters of stone. Many of the bluebloods from the industry are buried nearby—Ingalls, Elliot, Perry, Matthews, Reed—whose names are engraved in ritzier stones, in granite and marble. Near the foot of the stone cutter's pedestal is the monument for Louis Baker, sculpted to represent a carver's bench with a section of pediment on top and all the tools of the trade—calipers, square, chisels, hammer—scattered about. As you drift toward the oldest part of the cemetery, where the dates of birth run back before the Revolution, you leave marble and granite behind and enter a forest of limestone. The most eloquent of all these markers is a raw monolith, uncut, undated, tilting as if toward sleep and bearing only the faint initials "T. W." There is a primordial feel to this rough pointing finger of stone, as if it rose from the very roots of grief and praise. And over it all—over monolith and Grecian temples, over carvings of cupids and hound dogs and doughboys, over peony bushes, marigolds, and begonias, over plastic flowers and the lords of industry—looms the jaunty figure of the cutter, frozen in stone at the peak of his pride.

12 The Shape of Things to Come

In East Oolitic, on a street of drowsy houses, a green sign whispers:

QUIET
SICKNESS.

Reading this, you might be tempted to understand it not as a plea on behalf of an invalid in the neighborhood, but as the name of an epidemic disease. You might wonder if the sign speaks to the condition of all the stone towns, as the seas of prosperity withdraw, the noise of channeler and diamond saw dwindles, and a terminal quietness settles on these forgotten places.

In a country scarred with overgrown quarries and littered with the ruins of mills, a country inhabited by hundreds of old men who remember the sizzle and smoke of an earlier day, it is easy to feel that all the glories of limestone are past. In such country, nostalgia is a handy liquor. Like a traveler in Rome, overwhelmed by relics, you might not realize that all around you people are still living, still working, still making history. To be sure, the combination of forces that raised this lovely stone into prominence between 1890 and 1930—the building of the great northern industrial cities, the flood of immigrants, the spread of the railroads, the invention of the skyscraper, the revival of classical architecture—will never recur. But is the quarrying and cutting of limestone a doomed art, like horseshoeing, like the chipping of flint arrowheads? In an age of synthetics and throwaway architecture, is there a future for limestone?

"There's a healthy future, a bright one, if you know what you're doing and you turn out a quality product," Wilbur Bybee inists. "So long as anybody's putting up buildings they care about—churches, schools, banks, hotels, courthouses—they'll want to use stone." A sad-eyed and stout-bodied man in his fifties, Bybee pays out his words stingily, in a Hoosier slur. Five years ago, he demonstrated his confidence in that future by taking over the idle Matthews Brothers Mill in Ellettsville—a mill that had operated in the same location from 1862 until the late 1970s—and establishing the Bybee Stone Company.

"My three sons wanted to go into the stone business, so I needed a place for them to run while I'm still around and to keep running when I'm gone." They hired the best men they could find (including Matthews veterans), fixed up the antique machines (one of the diamond saws, still operating, was patented in 1870), bought some new equipment (in the office an Apple computer winks and flickers next to the woodstove), opened up the neighboring quarry (whose stone had won a prize in the World's Fair of 1876), and started bidding jobs. "Pretty soon jobs came our way, because this mill's always had a reputation for doing good work, especially fancy cut work." The old Matthews Brothers operation had shipped four hundred carloads of stone for the Baltimore Cathedral, and since 1962 had been cutting all the stone for the National Cathedral in Washington. Bybee and his sons revived that tradition of fine, ornate worksmanship, and as a result their mill is flourishing. "There's been some people in this business that would ship out some awful sorry stone. My view is you've got to take pride in what you ship. It's got to be just so. It's got to be right. When the builders put our stone on a building, I want it to fit like a key in a lock."

At any given time there might be twenty jobs going through the mill, ranging from single cornerstones or free-standing sculptures to entire buildings rolling away in pieces on a hundred trucks. Bybee runs down the current list: office towers in Detroit, classrooms at the University of Syracuse, an insurance company in New Jersey, a hospital in Maine, condominiums in Kentucky, an addition to the Smithsonian Institution in Washington, a natural resources building in Montana, a Baptist medical center in Tennessee, restorations on the Iowa State Capitol and the National Capitol in D.C.

"There'll always be restoration work. You just think about it, there's thousands of limestone buildings scattered all over America, most of them built with stone from these two counties. These are solid buildings, handsome buildings, too good for tearing down. So when people want fresh stone for repairs or additions, right here's where they'll come to get it."

What does he think about the trend in some other mills away from ornate work—the elaborately cut stones that jigsaw together to form the walls of buildings—toward large blank panels? "I don't think much of it, to be honest with you. There's money to be made there, I suppose. But slicing out nine hundred identical slabs and bolting them to a steel frame is not my idea of what limestone is for. I'll stick with the old way."

Inside the mill, where Bybee goes to inspect a gigantic mortar and pestle being carved for the headquarters of a drugstore chain, there is a balletlike precision to the movement of men and machines. On a tool cabinet nearby, someone has penciled a rhyme:

> To Do Is To Be   —Socrates
> To Be Is To Do   —Jean Paul Sartre
> Do Be Do Be Do   —Frank Sinatra

Bybee runs his palm over the lip of the mortar, says a word to the man who has been rubbing it smooth with emery cloth, then returns to the office. The man rubs harder.

"The stone business has been run for a century by a handful of families," says F. G. Summitt, cousin to the proprietor of Summitt's Grocery and Post Office, "and as the companies have been handed down through the generations, the new owners have failed to keep up with trends in architecture or construction or engineering. They're too conservative, too backward-looking. They want to cling to the methods and styles of fifty years ago." Summitt himself was born into that limestone aristocracy forty-odd years ago. His father, grandfather, great-grandfather, and sundry uncles were all stone men. But everything about him—from his dapper business suits to his impeccable speech, from his oak-and-chrome furniture to the sped-up tempo of his mill—marks him out as belonging to a new breed. There are university degrees behind his name, including an M.B.A. In his office, copies of *Progressive Architecture* and *Architectural Digest* cover the glass-topped coffee table. A blown-up color photograph of the Biltmore mansion hangs on one of the walls. On his desk, rather than sample lumps of limestone, there is a vase filled with delicately carved wooden flowers. Instead of trapping foxes or fishing for bass in his off hours, he jets around the country to watch sprint-car races. Instead of sitting tight in the stone belt and firing bids to cities far away, he flies to those cities and talks with investors and architects in their own arcane languages. He looks for ideas, not to his competitors in the stone belt, but to avant-garde marble and granite mills on the East Coast and in Europe.

Looking at Jeff's photographs, Summitt admires the art but objects to what he calls the sense of oppression and defeat. "What I see here are images of the past, of history. I respect what's gone before, but don't identify with it. That has nothing to do with where I'm going."

Wherever he is going, he appears to be well launched and moving swiftly. Slim and angular, a natty dresser, articulate but reserved, he radiates a sense of mission. Whether showing us blueprints in his office, sitting with us on the dry stone bed of a creek to eat a delicatessen salad, or tramping along a railway spur and leaping over barbed-wire fences in search of an old quarry, he remains formal and urbane, without a whiff of the backroads about him. (Being roughly seventy percent backroads myself, I can detect the presence of country in other people at levels of a few parts per thousand.) His talk is carefully measured, premeditated, lacking not only the profanity but also the grammatical inventiveness so common among stone men.

"I think limestone has unlimited potential. Most of the industry is antiquated, that's true. Many companies are doomed, because the people who run them cannot imagine changing the way they do things. But this is a fact about certain businesses, not about the material. The stone itself is superb, an extraordinary natural product. The challenge is to fashion it in a way that customers want."

For a little more than ten years, Summitt and his partner, Larry Evans, have been fashioning limestone into flat wall panels, which fit together like children's construction blocks to form skins around the steel skeletons of buildings. Roughly four inches thick, these panels come in modular sizes (ranging up to five by fourteen feet), standard colors (such as Cathedral Rizzo and Palestine Goldentone), and standard finishes (including Shantung and Sugarcube). Customers from the Carolinas to California have been ordering them at a brisk and accelerating rate. With their business doubling every two or three years, Summitt and Evans keep a fleet of trucks busy hauling stone from their two mills in Bedford. One of the mills is devoted exclusively to the fabrication of wall units and spandrels out of stone, steel, and epoxy. The other one looks very much like a traditional cut-stone plant, except that all the machines are running faster, the equipment is newer and in better repair, the men are younger and more alert.

"You notice how few of them chew tobacco?" Summitt points out during our tour. With mouths empty of tobacco but full of teeth, they look fitter than the men we've seen in other mills, without the customary guts hanging over their belts. They have about them the crisp, cocky air of jockeys riding the lead horse. Beside the clock-card alley, where they must pass both coming to work and going home, a glass case displays photographs of recent Summitt and Evans jobs—bold, clean-lined insurance buildings, hospitals, synagogues, schools, banks, museums—in Miami, Dallas, Cheyenne, Joliet, Utica, Baton Rouge, Louisville, Memphis. These pictures, and the company's glossy brochures, remind the men what their labor adds up to.

Everywhere in the mill, the tinkering hand of engineers shows in small improvements: the conveyors are motor-driven, the gang saws rock at double speed, the beds on the planers are wider than customary, and the cutting tools are broader. There is little dust, because most of the cutting operations are sluiced down with water. In a new wing added along one side of the mill, an ingenious assembly line has been constructed for cutting large numbers of uniform pieces. Raw slabs enter at one end, pass through a dozen grinding and sawing and polishing devices, and emerge from the far end as finished panels ready for setting in a wall. The four or five men who tend this mechanical stream accomplish in a day what, by the old methods, would have occupied a dozen men for a week. "I realize this means transferring skills from men to machines," Summitt concedes. "I realize there is no place in our operation for carvers, and very little work for cutters. But that is the inevitable trend for all industry. Hand craftsmanship, much as I admire it, cannot produce a modern building material."

Summitt's admiration for the old ways of quarrying and carving stone is informed by a lifetime of study. Born and reared in Ellettsville, he used to walk the railroad tracks and creekbeds leading out from town, a gunnysack over his shoulder for gathering hickory nuts, eyes peeled for the gray shoulders of stone. He climbed down into every quarry hole he could find, and sometimes had to shinny up a tree in order to climb back out. Still today he searches out the sites of quarries and mills, asking old-timers for directions, following maps and hints in out-of-print books. (He keeps a standing order at all the local bookstores for anything to do with limestone.) Riding with him through the stone belt is like accompanying an archaeologist through a dig: he sees on all sides the layered history of this place and its people. There is a tension in him, as there must be a tension in anyone who spans the gap between a past era and a future one. Of all the people we've met, he knows the most about the old ways, about the days of ox-cart and mallet-and-chisel, and yet at the same time he is the most firmly committed to the pursuit of new ways. "I love carving, but I'm willing to let it go. I respect the men who built gothic churches, the men who laid out vaulted ceilings, the men who cut and raised the intricate towers. I've learned from them, but I feel the need to go beyond them. My deepest loyalty is to the stone itself. I'm determined to preserve limestone as a modern building material, not just a museum curiosity."

Wanting to show us Little Giant Quarry, abandoned a century ago, Summitt leads us jouncing one day half a mile up a rutted drive to a house built of rough-sawn lumber. He tells Jeff to honk the horn, so the folks inside won't think we're sneaking up on them. Then he jumps out, braving dogs and possible shotguns, and asks the skittery couple who meet him on the porch if we may look at their quarry. They blink at him—a man wearing European sunglasses; long-sleeved, button-down, oxford-cloth shirt; thirty dollar haircut; spotless chinos with knife-sharp creases—and they blink again, but say yes. Although they have lived within a stone's throw of the quarry for eight years, no one before Summitt has ever told them what it's called. In the overgrown stacking yard of Little Giant, nimble and curious, he leaps from block to block, chipping off a corner here and there to judge the quality of the stone. Poking around, he speculates on how long the quarry operated, why it failed.

Summitt has no intention of failing. "Right now we are state-of-the-art for limestone. What we're doing is very successful. We're so smothered in orders we can barely breathe. It's tempting to just keep doing what we're doing. But that's what has doomed so many outfits in this region. Already we're looking toward new products, toward the *next* stage in building and architecture." He speaks of attaching even thinner veneers, perhaps no more than one inch thick, to steel frames. Insulated, drilled for pipes and wires, these modular units would then supply both skin and skeleton for buildings of the future. "You would bolt them together, screw gypsum board to the inside, and be ready to hang your pictures. That is the future of limestone. Not rose windows. Not archways and gargoyles. Whoever doubts that can sit and wait, and go under."

"Without question, that's where things are headed, toward wall panels as skinny as books and as blank as the faces of cows," says Bill McDonald, head of the Indiana Limestone Institute, the industry's trade association. "But I don't think we should go whole-hog in that direction. No matter what new building technologies we get into, there's always going to be some of the traditional business, the pierced tracery and fluted columns. We shouldn't turn our backs on our history." McDonald has been part of that history since 1960, when he quit selling mirrors (the company he worked for went bankrupt) and started selling limestone. He's midway through his fifties, sandy haired, with active eyebrows and a spry little sandy moustache that accentuates a mobile mouth. He is a talker. He talks as easily and uproariously as a circus ringmaster. In his office on the top floor of Bedford's Stone City Bank, he pumps the air full of words, chopping them into sentences with great sweeps of his meat-cleaver hands, leaning forward and rocking backward, kicking his legs for emphasis, an athletic talker who gives himself quite a workout without stirring from his chair.

"If we're going to make panels, I hope to high heaven we don't slice the veneer too thin, or we'll have limestone flying around the cities. That's been a problem with the synthetic materials—glass, terra cotta, poured concrete, aluminum, you name it—but never limestone. Limestone is the best building material God ever put on earth. It doesn't rot, warp, peel, rust, shrink, crack, check, or jump out of the wall. It just sits there. And that's the first thing you want a building to do—just sit there."

McDonald gives the language quite a workout, too, veering in the space of a sentence from the abstractions of modernist aesthetics to Anglo-Saxon oaths that are remarkable for their concreteness: "These architects who throw up buildings as art objects only—never caring if they work as *buildings*—just shit in their own messkit." One of his favorite epithets, for describing whatever he regards as foolish, is "dumb-assery." "I've given up being a curmudgeon, but I'm still a Cassandra. When I see dumb-assery in the design of buildings, I don't mind saying so. Politely, of course, since I make my living by dealing with architects."

Each month he mails out literature to over ten thousand architects and builders, espousing the cause of limestone. He tours the country, giving seminars, showing films, lecturing to classes, pouring the oil of talk into every available ear. "Those are my only tools, you see—a head full of knowledge and a whole truckload of talk." A lot of what he has seen during his quarter century of traveling disturbs him. "We're just now leaving behind a horrendous period in architecture. You look at buildings put up in the sixties and seventies, and there's a Kleenex feeling about them: use them once and throw them away. Engineers who worried only about saving bucks squeezed all the slop out of buildings—all the safety factors—and windows began to pop out, joints cracked, façades rained down on the streets, the west sides of buildings cooked while the east sides froze. Architects tried out every gimmick that came along, just for the sake of novelty. Nowadays you could make a building out of crushed onions if you wanted to. The technology exists. Just glue it together and stick it up. But would you want to? If you ask me, people are sick of gimmicks and cheapies."

McDonald blames Hitler for many of these follies, because the Bauhaus architects who fled Nazi Germany settled in America, and within a generation they and their students transformed our vision of buildings. "Ornament is crime, they taught. What they gave us were spare, glassy, skinny-membered art objects, many of which are god-awful to live or work in, many of which—we now discover—are maintenance nightmares. You can't heat them, can't cool them, can't keep the rain out. The problem is, nature never went to architectural school. The forces of gravity are absolutely implacable. And so is the weather. If you cut out all the slop, all the margin for error, eventually the forces of nature will catch up with you. This is why masonry buildings are returning to favor, and why limestone in particular is enjoying a renaissance." Again he tells us (as with a real ringmaster, McDonald gives me the feeling that successful lines from today's show will turn up again, verbatim, in tomorrow's), "Because, you see, limestone is the best building material God put on earth. It doesn't rot, warp, peel, rust, shrink, crack, check, or jump out of the wall. It just sits there. And that's the first thing you want a building to do—just sit there."

Our worries about energy are also contributing to a limestone renaissance. Compared to other building materials—such as aluminum, steel, glass, or brick—limestone requires little energy to produce, and because of its massiveness it serves as buffer between inside and outside, storing energy the way a flywheel does, damping the ups and downs in a day's temperature cycle. "When energy was cheap, people built pretty glass boxes and ignored the utility bills. But in 1973 the Arabs put an end to that era. With their boycott on oil, they suddenly made us all aware that energy was no longer going to be cheap. People began to realize once again that buildings must be good, efficient machines *first*, art objects second. And if you want to construct a tight, solid, energy-efficient building, you can't use anything better than limestone. Once again it's becoming a material that world-class architects—people like Philip Johnson and Michael

Graves—are eager to work with. And this makes limestone a premium material in the eyes of lesser folks. That's why the people in our business ought to wear burnooses and bow four times to the East every day, to thank the Arabs for bringing us all back to our senses."

At night, when he has laid aside his metaphorical burnoose and his literal shirt and tie, McDonald writes fiction. "Strictly adventure tales and thrillers. I'm not a deep thinker on anything except limestone. When I leave work, I'm content to think frothily." One day he means to get his novels published, but in the meantime he keeps pecking away at the keyboard, night after night. He writes in the evenings because his brain balks in the mornings. "I'm a slow riser. When I first get up I run into walls. You know, I really wonder about these people who wake up with a start and are raring to go. They're atavists, throwbacks, hairy-knuckle types. Deep down in their blood they're still afraid of saber-toothed tigers. We late risers are a more modern branch of the human tree." A thought occurs to him. His rusty eyebrows draw together, lift cautiously. "Say," he asks me, "when do you write?"

"From six A.M. until noon," I answer.

"It just goes to show. There's no generalizing."

In fact McDonald loves to generalize. During the three hours of our talk he leaps gleefully into the midst of every subject—religion, music, landscape, education—and plays his opinions in concertos of words. But no matter where the performance begins, it always returns, inevitably, to his obsessive motif, the praise of limestone. "There is no material, no industry, quite like this one. Here you have this magnificent natural stuff located in a tiny pocket of hills in southern Indiana. Bucolic southern Indiana. And most of the people who work with it are not sophisticated people. Smart people; good people—but not sophisticated. Over the years a lot of companies have been run like mom-and-pop groceries, like family farms. We live in one another's back yards. We know whose check is good and whose wife isn't. It's unkind, perhaps, but it's not untrue to say that most of the people in the industry ship their stone off into a great void, and thirty days later, out of that void, back comes a check. And yet the material they sell has always been used and is still being used for some of the most significant and innovative buildings in America. You see, it's the stone itself that's the key. The industry will go on, thriving at times—like right now—and suffering at times, expanding and contracting like an accordion. But it will always come back. We're survivors. Not because we're noble or wise. We're resilient, sure. But the key is, we have under our feet the best building material God ever put on earth. Because of that fact, this industry is as nearly eternal as you can get. One hundred years from now, people will still be hauling limestone out of this little patch of ground. They may be shipping it on spaceships or light rays, but one way or another they'll be hauling it out of the ground and stacking it into the air."

Epilogue
In Praise of Limestone

Wherever holes have been drilled in quarry ledges or water has licked open the seams in outcrops, dirt catches and seedlings take root. Eventually these roots will burst the stone. Our own roots also go down into rock—the rock of caves, spearheads, knives, the megaliths and cairns and dolmens of our ancestors, the rock of temples and pyramids, gravestones, hearthstones, cathedrals. Entire millennia of human labors are known to us solely through their stone leavings. The only common stuff that rivals it for durability is language, words laid down in books and scrolls like so many fossils. With a touch of mind, the fossil words spring to life; so might the stones, if we look at them aright.

While you read this, new limestone is forming on the sea floor near the Bahamas, grain by grain, corpse by corpse. You could dive down and grab it by the fistful, doughy still from the creation. It's also dissolving away right this moment in hidden caverns beneath my feet. As W. H. Auden wrote in a poem about the limestone coun-

In Praise of Limestone

tryside of his native Yorkshire:

> If it form the one landscape that we, the inconstant ones
> Are consistently homesick for, this is chiefly
> Because it dissolves in water.

Rain and melted snow seek out every fault, scour passage through every fracture, open tunnels and sinkholes, underground rivers and sudden gurgling springs. Water makes it, and water unmakes it.

But the laying up and the wearing down of limestone is slower than the creep of glaciers, slower than the thickening of trees, so very laggard and stately it makes everything human seem hasty. It reminds me that, however temporary we two-leggers may be, earth is no flash in the pan. Not only our hearts and minds are inconstant, but so likewise are the bodies and cities and nations we inhabit. Most of what we manufacture these days is less durable than a dayfly. New streets buckle in the first frost, chipboard walls sag under the weight of pictures, cars rust through before the last note is paid. At night we lie awake listening to the motors moaning and gears snapping in our housefuls of appliances. Governments rise and fall like yo-yos. At a time when our own lethal inventions could erase within the space of an hour everything our species has ever made, when the poisons we spew into air and water begin to snuff us out, when our multiplying mouths threaten to eat us out of house and planet, I find keen comfort in the enduringness of limestone.

I am glad to live in this pocket of rumpled hills where the crust of earth shows through. When the fog of human voices grows too thick for my lungs, and the ticking of my own inner clock rattles my soul, and I feel the winds of momentariness whistling through my ribs, I go out to climb a cliff or splash down a stony creekbed or dangle my legs over a quarry's lip. Some future day, oceans will wash again over this spot where I sit in my rickety chair, and once again myriads of beasties, many of them indistinguishable from the sea creatures of three hundred million years ago, will spawn and die in the shallows of Indiana.

Words have become so cheap, millions are flung at our ears and eyes every hour. The words I lay down this minute are no more than pinpricks of light on a dark screen, fragments of language settling like bits of shell in the warm waters of a tiny sea. On the table beside me as I write, I keep within easy reach a rectangular slice of limestone. It has taken on a rain-dark cast from the oil of my palms. Its regular shape speaks to me of those who dig and cut, reminds me to honor the work of hands. The body of the stone itself, a cake of gauzy webs and nebulae and spiral galaxies, speaks to me of the force that curls through my fingers and drives the tides and sets the wondrous worlds to spinning.

Part Two: THE STONE BELT

1 The Land and Its Transformation

Landscape, Oolitic, 1983

Waterfall, McCormick's Creek, 1984

Harrodsburg roadcut, State Road 37, 1984

Overlooking Empire State Quarry, Oolitic, 1983

Cottonwoods with spall, Woolery Stone Co., 1983

Grout piles, University Quarry, 1983

Derricks, Woolery Stone Co., 1982

Statehouse Quarry, McCormick's Creek, 1984

Empire State Quarry, Oolitic, 1983

Car with bullet holes, Oolitic, 1984

PCB contamination, Bennett Quarry, Bloomington, 1984

Trestle, Victor Pike, 1984

Cedrick Walden, retired quarry worker, Stinesville, 1984

Nineteenth-century quarry, near Stinesville, 1984

Emmett Hoene with family, Woolery Stone Co., 1983

Betsy in Long Hole, Sanders, 1983

2 Quarrying

Last day of quarry season, Independent Limestone Co., 1983

Independent Limestone Co., 1984

Winter, Maple Hill Quarry, 1983

Mud tracks, Reed Quarry, 1984

Tree root, Reed Quarry, 1984

Night, Maple Hill Quarry, 1984

Reed Quarry, 1983

Turning a cut, Ledge F, P.M. & B. Quarry, Oolitic, 1983

Elliott Stone Co., Eureka, 1984

Breakers, Maple Hill Quarry, 1983

Blockmarker, B. G. Hoadley Quarry, Guthrie, 1983

Reed Quarry, 1984

Joe Wilson, Adams Quarry, 1983

John Smith, pump man, Independent Limestone Co., 1984

Gene Maddox and Stanley Phillips, B. G. Hoadley Quarry, Guthrie, 1984

Night, Ledge F, P. M. & B. Quarry, Oolitic, 1983

3 Stone Mills

Jimmy Leach, Edinger Stone Co., Bedford, 1983

Stencils with map of Indiana, Woolery Stone Co., 1982

"Art Gallery," Woolery Stone Co., 1981

Cucumbers with sports page, Woolery Stone Co., 1982

Donnie Martin, Bob Woolery, and Emmett Hoene, Woolery Stone Co., 1982

Bybee Stone Co., Ellettsville, 1984

Ernest Hattabaugh, gang sawyer, Woolery Stone Co., 1982

Delbert Willison and Birtle Hawkins, hookers, Fluck Cut Stone Co., 1983

Madonna, Woolery Stone Co., 1982

Pheasant, Woolery Stone Co., 1982

Roy Hamm, Bybee Stone Co., Ellettsville, 1984

Looking west, Woolery Stone Co., 1982

Bybee Stone Co., Ellettsville, 1984

Gang saw blades, Woolery Stone Co., 1982

Marvin Brinson, diamond sawyer, Woolery Stone Co., 1982

Woolery Stone Co., 1984

Jerry Sowders, Woolery Stone Co., 1982

Henry Morris, stone carver, 1984

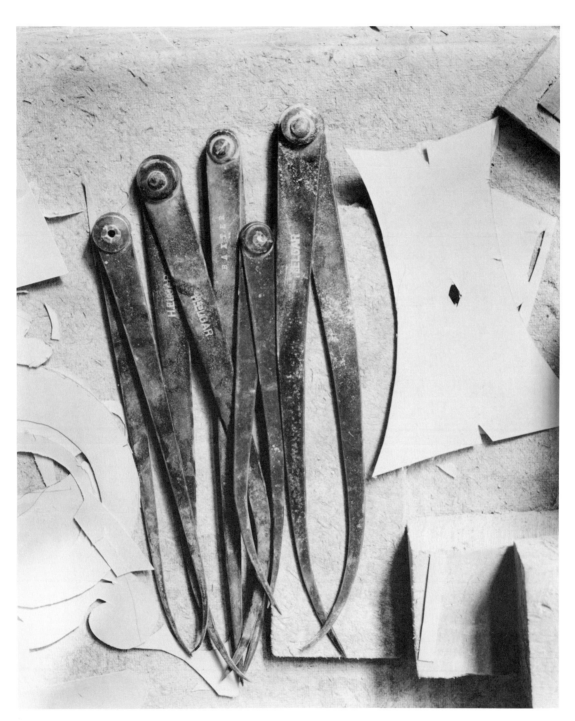

Henry Morris's calipers, 1983

Light, Bybee Stone Co., Ellettsville, 1983

4 Works

Marker, Ellettsville, 1984

Limestone Tourist Center, Oolitic, 1984

War Memorial, Indianapolis, 1984

Flatiron Building, New York, 1984

Empire State Building, New York, 1984

Grave marker, Oolitic, 1984

Old City Hall, Bloomington, 1984

Playboy Building (formerly Palmolive Building), Chicago, 1984

Central Business College, Indianapolis, 1984

Stone masons, National Cathedral, Washington, D.C., 1984

Rockefeller Center, New York, 1984

American United Life, Indianapolis, 1984

Colonnade, Union Station, Chicago, 1984

Chicago Public Library, 1984

Metropolitan Museum of Art, New York, 1984

Frog, Woolery Stone Co., 1984

Roof, Ellis Island, New York, 1984

Hieroglyphics from Saqqara, Egypt (limestone carving c. 2400 B.C.)

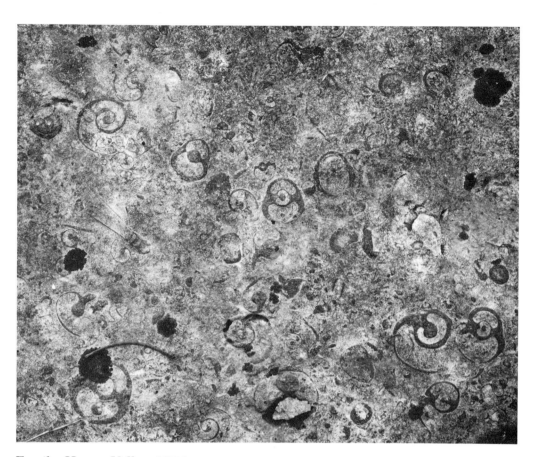

Fossils, Hunter Valley, 1984

Afterword

For three years I have been exploring photographically the region in which I live. I wanted to see if I could find out how the land influenced people, how experience of a place might be photographed and ultimately provide me with insight and understanding. I didn't start out to make a book of photographs (that idea occurred more recently, when the photographs themselves suggested their grouping together in a book), but to gather some thoughts, some feelings, some impressions on film. Earlier work, things I had done before moving to the hills of southern Indiana, looked at the relationship of man and nature; the work I did here in the stone belt continued this sense of ecology.

Stone quarrying is an extractive industry and reflects man's idea that everything on this earth is here for us to use. Rolling hills are cut and sliced and shipped all over the country. The resultant landscape is a chaotic jumble of gaping holes and rusting steel. Quarries fill with water and become swimming holes; one quarry area in Bloomington was considered among the best swimming spots until it was found to be contaminated with deadly PCBs, dumped from the local Westinghouse plant. People drive cars into abandoned quarries and later use the auto bodies for target practice with their guns. As Scott Sanders points out, the quarries are places of violent activity; I suppose I see them as a kind of elemental battleground—an interface where man wages war upon nature.

There is, however, another side to be looked at. For nearly every hole in our backyard here in the stone belt (the serpentine outcrop of building-grade limestone that stretches from Bloomington south to Bedford, twenty miles away), there is a building or monument somewhere. Something like three-quarters of all stone buildings in the United States are made from our local stone. It is intriguing to identify some huge hole in the ground as mother to the Empire State Building. I have followed chunks of limestone ripped from the ground at Independent Limestone Co. to the towers of the National Cathedral (with a brief layover at Bybee Stone Mill).

Nature ultimately triumphs; given enough time the land begins to revert to a more natural state and man's presence begins to disappear. The more recently introduced PCBs notwithstanding, the old quarries themselves are relatively benign, new habitats for plants and animals. And the scale of this industry is relatively small, infinitesimal in comparison to, say, coal strip-mining.

Another aspect of the photographs included in the book concerns portraits of men who work with limestone in quarries and mills. These are blue-collar workers whose fathers and fathers' fathers worked with limestone in the stone belt. They now face the uncertainty of an industry (perhaps the oldest one, tracing its roots back to the stone age) that finds itself in a downward spiral in this era of high technology and disposable architecture. I wanted to look through their eyes to gain insight into their ideas and feelings about the land, their back-breaking jobs, our precarious times—in short, their views of the world.

In the process of making the photographs that appear in this book and in my wanderings all around the stone belt and cities where the limestone now resides, I used a variety of camera formats. The majority of the images, however, were made with an 8" × 10" view camera—something of a fossil of a camera in the age of electronics and somehow appropriate for photographing an antiquated industry that deals in fossils. The prints from 8" × 10" negatives are made by contacting them onto photographic paper without enlargement. This insures that the images contain a wealth of photographic detail and integrity of tone unattainable by any other means.

The photographs in this book are not to be seen as objective documents. Rather they are intended as personal responses, aspiring toward the poetic, to some thoughts and feelings, some experiences I've had in my travels in the stone belt.

JEFFREY A. WOLIN
BLOOMINGTON, 1985

Editor: Risë Williamson
Book designer: Matt Williamson
Jacket designer: Matt Williamson
Production coordinator: Harriet Curry
Typeface: Bodoni Book
Typesetter: G&S Typesetters, Inc.
Printer: Malloy Lithographing, Inc.
Paper: 70 lb. Glatco matte coated
Binder: John Dekker & Sons, Inc.
Cover material: Holliston Kingston natural